Reductionism

A Beginner's Guide

ONEWORLD BEGINNER'S GUIDES combine an original, inventive, and engaging approach with expert analysis on subjects ranging from art and history to religion and politics, and everything in-between. Innovative and affordable, books in the series are perfect for anyone curious about the way the world works and the big ideas of our time.

aesthetics
africa
american politics
anarchism
animal behaviour
anthropology
anti-capitalism
aquinas
art
artificial intelligence
the bahai faith
the beat generation
the bible
biodiversity
bioterror & biowarfare
the brain
british politics
the Buddha
cancer
censorship
christianity
civil liberties
classical music
climate change
cloning
cold war
conservation
crimes against humanity
criminal psychology
critical thinking
daoism
democracy
descartes
dewey
dyslexia
energy
the enlightenment
engineering
epistemology
european union
evolution
evolutionary psychology
existentialism
fair trade
feminism
forensic science
french literature
french revolution
genetics
global terrorism
hinduism
history of science
homer
humanism
huxley
iran
islamic philosophy
islamic veil
journalism
judaism
lacan
life in the universe
literary theory
machiavelli
mafia & organized crime
magic
marx
medieval philosophy
middle east
modern slavery
NATO
nietzsche
the northern ireland conflict
nutrition
oil
opera
the palestine–israeli conflict
particle physics
paul
philosophy
philosophy of mind
philosophy of religion
philosophy of science
planet earth
postmodernism
psychology
quantum physics
the qur'an
racism
reductionism
religion
renaissance art
shakespeare
the small arms trade
sufism
the torah
united nations
volcanoes

Reductionism

A Beginner's Guide

Alastair I. M. Rae

ONEWORLD

A Oneworld Paperback Original

Published by Oneworld Publications 2013

A CIP record for this title is available
from the British Library

ISBN 978-1-78074-254-0
eISBN 978-1-78074-255-7

Typeset by Cenveo Publishing Services, India
Printed and bound by Nørhaven A/S, Denmark

Oneworld Publications
10 Bloomsbury Street
London
WC1B 3SR

To David

Contents

Preface

Scottie, the spaceship engineer in *Star Trek*, was often heard to tell the captain 'Ye cannae break the laws of physics, Jim!' What are the implications of this statement, which most people believe to be true? The laws of physics determine the properties and behaviours of the atoms and other particles that are the fundamental building blocks of matter. When atoms come together they form molecules and when large numbers of molecules are joined up solids and liquids are produced. Are the scientific laws that apply to, say, liquid water and solid ice the same laws of physics that govern the behaviour of the component H_2O molecules? Reductionism says that these fundamental laws are indeed all that are needed; large-scale properties, such as the fact that water is a liquid and ice is a solid, emerge as consequences of the way their molecules are arranged and of the interactions between them.

So far, so uncontroversial but suppose we apply the same reasoning to biological systems. Is it really true that the complex structures of biological cells, with their DNA and wide variety of protein molecules, emerge when the basic laws of physics are applied to their constituents? What about properties of the human brain and mind, such as consciousness? Do these also emerge in a similar way or are they controlled, at least to some extent, by fundamental forces whose origin is outside physics? What about when human beings come together to form societies: can the laws of sociology and economics be reduced to those governing the behaviour of individuals?

Some readers will answer 'yes' to all these questions, while others will disagree. Either way, many will have only a general, perhaps even vague, understanding of how science accounts for the connections between these different phenomena. In this book, I have tried to provide an account of how reductionism applies at every level, from the atom to the behaviour of human societies, and to make this accessible to a general readership. As my professional discipline is physics, I have had to venture outside my comfort zone to include subjects such as biology and economics. One consequence of this and the wide field of knowledge covered is that more specialist readers may find some sections rather elementary. If so, I hope that they will still appreciate seeing how their specialism relates to my overall theme and, even if they think the discussion is over-simplified in places, that they will not find it actually wrong. If they do, the responsibility is, of course, all mine.

Acknowledgements

I am grateful to my former students and colleagues of the University of Birmingham, where I taught for over thirty years, for their patience and inspiration. I am very grateful to Peter Main and Ted Forgan for their penetrating and constructive criticism of large sections of the draft, as well as for stimulating conversations. I thank my son Gavin for his guidance on economics and politics. I should also like to thank my editor, Robin Denis, for providing continual help and encouragement; the copy editor, Ann Grand, for her careful reading and close criticism of the draft text; and others at Oneworld Publications. I gratefully acknowledge personal help I received from Maria and John in 2012.

My wife, Ann, continues to tolerate and encourage me and her support has played a vital part in ensuring the completion of this project, as did that from my two sons and three grandchildren.

1

I reduce

Pigs might fly. This phrase is commonly used to describe something that is impossible and will never happen. But what would we conclude if we looked up into the sky and actually saw a pig flying past? Our first thought might be that it is a lighter-than-air balloon in the shape of a pig or that there must be some other 'natural' explanation. Why do we believe that flying pigs (and indeed flying carpets and broomsticks) are impossible, while objects such as most birds, some insects, aircraft, and rockets do indeed fly? The scientific answer is that all objects are subject to the law of gravity but the latter entities are constructed so that this force is countered by their ability to flap wings, spin propellers, or emit powerful exhausts. The first of these statements follows from the fact that gravity is a universal force that acts on the bodies' component atoms independently of how they are arranged to create their external shapes. This is an example of *reductionism*, the fundamental principle that underlies all scientific reasoning and is the main subject of this book. Put simply, it states that the fundamental physical laws governing natural phenomena are the same as those applying to their basic constituents. Thus, for example, the total gravitational force on an object is the sum of the forces acting on the individual atoms, whether the object is a stone, a pig, a carpet, or even an aeroplane.

The principle of reductionism plays an important role in the scientific process, forming a central, if often unstated, assumption underlying almost every scientific statement. Reductionism means that the properties of something being investigated can be understood as emerging from the properties of its component parts and an understanding of how they interact. Thus, a biologist may explain a phenomenon such as the movement of a limb following a muscle contraction in terms of the properties of the biological cells that form the muscle – so 'reducing' the limb movement to the behaviour of its component cells. A biochemist may study a typical muscle cell and describe it in terms of the properties and behaviour of the molecules that make it up, hence 'reducing' the cell properties to those of the molecules. A chemist may examine one of the molecules and find that it is composed of atoms, and an atomic physicist might describe an atom as composed of a nucleus surrounded by a number of electrons which obey the laws of quantum physics. In this way, it could be said that biology can be reduced to atomic physics. Indeed, the physics of the nucleus and electrons can be further reduced to that of its constituent particles and these, in turn, emerge from the laws of what is known as quantum field theory. Beyond that some believe (or hope) that there is a 'theory of everything' waiting to be discovered. Attempting to understand what is happening at this most basic level is a major aim of some scientists and has motivated the building of the multibillion-pound Large Hadron Collider in Switzerland, where a long-predicted but never previously detected particle known as the 'Higgs boson' was discovered in 2012.

Further applications of reductionism relate to the behaviour of human beings as individuals and to society as a whole. Can our thinking can be reduced to the properties of our brain and can social behaviour be reduced to that of individuals? I shall be discussing these sometimes highly contentious ideas, as well as the less controversial applications of reductionism to the physical and biological sciences. Three important ideas form a tool-kit that will be used in tackling this task; these are the principles of *falsification*, *simplicity*, and *emergence*.

Falsification

What would be the logical conclusion to draw if we actually saw a flying pig, assuming that all natural explanations – such as it being a lighter-than-air facsimile – have been eliminated? The only alternative would be to conclude that the law of gravity could not have been acting on the pig at the place and time we observed it, so there must be something wrong with the laws of physics as they are presently understood. This would be an example of the principle of *falsification*, where a general proposition (the universal law of gravity) is shown to be false if one of its consequences (the impossibility of flying pigs) is found to be untrue. A classic illustration of falsification, from the world of biology, is the statement 'all swans are white'. For centuries Europeans believed this to be true, because every swan they had ever observed was white. Soon after Europeans discovered Australia, however, they encountered black swans, which immediately disproved the general proposition.

Modern science, in its efforts to understand the universe and the processes underlying it, embraces the principle of falsification. It is often believed that science has developed theories which truly describe many aspects at least of the physical world. There is no way, however, that this can be proved absolutely because, however much evidence is found that supports a theory, it is always possible that it will be falsified by some future observation.

The question of how a general conclusion can be drawn from a series of repeated observations is known as the 'problem of induction' and has been studied over the ages. The principle of falsification was introduced into this debate by the philosopher Karl Popper. Born in Austria in 1902, into a family that had converted to Lutheranism from Judaism, Popper worked in Austria until 1937, when he became a refugee from Nazism. After a stay in New Zealand, he settled in London in 1946, where he remained until his death in 1994. Popper made a number of significant, though sometimes controversial, contributions to philosophy but he is probably best known for his contribution to the philosophy of science, which is set out in *The Logic of Scientific Discovery*. First published in 1934, this work proposed that the problem of induction could be resolved by emphasizing the role of falsification. In Popper's view, the purpose of a scientific investigation is not to look for evidence that *supports* a theory but to carry out experiments that might *disprove* it. Thus, at any stage in the development of the scientific understanding of a physical phenomenon, there is a provisional theory (such as 'all swans are white') which has not yet been disproved. When further observations are made (like visiting Australia and seeing a black swan) the results should be examined to see if they are consistent with the proposed theory. If they are not, a new theory has to be devised that explains the new result and also accounts for all the earlier observations that were consistent with the old theory. The new, more sophisticated theory is then accepted as true unless and until it is in turn disproved by further experiments. Popper summed this up in the aphorism 'Good tests kill flawed theories; we remain alive to guess again'.

Popper developed these ideas further to propose a definition of scientific knowledge: 'In so far as a scientific statement speaks about reality, it must be falsifiable; and in so far as it is not falsifiable, it does not speak about reality'. This means that to qualify as scientific, a statement's content must be falsifiable in principle. This does not mean that it has already been falsified – because then it would be known to be false – but there should always be a possible test which would falsify the statement if it came up with a particular result. For example, 'the sun rises every morning' would be falsified if one morning this did not happen. Similarly, 'pigs do not fly' could be tested by observing the behaviour of pigs. In contrast, a statement such as 'grass is always bright red unless someone is looking at it (either directly or indirectly), when it turns green' is unscientific because, whatever the reality, it could never be falsified.

As an example of how scientific understanding employs the falsification principle, consider how the theory of gravity evolved. Isaac Newton, who lived from 1642 until 1727, was allegedly inspired to propose his theory of gravity after observing an apple falling from a tree, showing that there had to be an unseen force attracting the apple to the Earth. He generalized this idea by postulating that this same force acted between any two massive objects, including astronomical bodies, and that this is the reason why the moon moves round the Earth and why the Earth and other planets move in regular orbits about the sun. He was able to express his ideas mathematically to make precise predictions of this orbital motion, which agreed with the results obtained by observations made by astronomers.

Newton's theory of gravitation held for around two hundred years, during which time increasingly precise measurements of planetary motion were made. The only observations that did not quite agree with Newton's predictions related to some fine details of the orbital motion of Mercury, the planet closest to the sun. Various proposals were made to explain this discrepancy without abandoning Newton's theory; these included the suggestion that another planet (provisionally named 'Vulcan') moved in an orbit that was even closer to the sun and that its gravitational force affected Mercury's motion. However, no direct evidence of Vulcan was found and, indeed, later observations have shown that no such planet exists.

Given the discrepancy between theory and experiment, a new theory was needed and this was produced in the early twentieth century by Albert Einstein. Einstein was born in Ulm (which is now part of Germany) into a non-practising Jewish family. His reputation was first established by three papers published in 1905, before he had acquired a position as a professional scientist. These papers made major contributions to several fields of physics, some of which we will return to in later chapters. Einstein is generally regarded as the greatest scientist of modern times and he has certainly acquired a fame that vastly outshines that of any possible rival. His greatest work is probably the *General Theory of Relativity*, which was published in 1916, when he was Director of the Kaiser Wilhelm Institute for Physics in Berlin. This set out an alternative theory to Newton's law of gravitation, which had been falsified by the observations made on the planet Mercury. Einstein's theory proposed that gravitational attraction is the consequence of the distortion of space itself by the presence of massive bodies. In nearly all practical circumstances, the predictions of the new theory are indistinguishable from those of Newton but in the case of the orbit of Mercury, Einstein's, rather than Newton's, predictions agree with the experimental observations. The general theory of relativity also predicted the results of some other experiments that had not been performed at that time. In particular, it predicted that the path followed by a light beam would bend as it passed through the gravitational influence of the sun. This was confirmed by a joint British-German expedition to observe a solar eclipse shortly after the end of the First World War.

Newton's theory was generally accepted until the discrepancy in Mercury's orbit was identified, which was correct, as all its previous predictions had been experimentally confirmed. So far, general relativity has survived all experimental tests, which include explanations of the behaviour of 'black holes', where the gravitational forces are so large that all matter has been crushed into a point. Theoretical difficulties, however, arise when general relativity is combined with quantum theory. At present, experimental study of situations where the quantum nature of gravitation is significant is not practicable, but if and when they are, both general relativity and quantum theory may well have to be replaced by an even more sophisticated theory. Until then, the general theory of relativity should be accepted as a true – but provisional – explanation of the motion of objects under gravity and the predictions of any future theory will have to agree with those of general relativity in all the areas where it has been successfully tested.

Simplicity

Another important principle that underlies any scientific explanation is that it should be as simple as possible. This idea is often attributed to William of Occam, a Franciscan friar who lived in Surrey in the fourteenth century. When Occam first proposed that 'entities must not be multiplied beyond necessity', this was primarily a theological statement connected to Occam's belief that the only fundamental entity was God. Today, however, this principle is applied to the scientific process. In this context, his statement, which is often referred to as *Occam's razor*, means that a good theory should involve no more assumptions (entities) than are necessary to explain all the known facts. As an example of a bad theory, one could postulate that the force of gravity depends not only on the mass of an object but also on its colour – perhaps that a red object would be subject to a stronger gravitational force than a blue one. As it has been observed that otherwise-identical red and blue objects fall to the ground in the same way and take the same time to do so, an additional force would have to be postulated to explain why the motion is exactly the same as in Newton's 'colour-free' theory. Occam's razor says that this more elaborate theory should be rejected because it contains additional entities – gravity plus another colour-dependent force – instead of gravity alone.

Another illustration is the development of our understanding of the solar system. It is often said that the first person to suggest that the sun rather than the Earth sits at the centre of the solar system was the Polish astronomer Copernicus, who worked in

Krakow early in the sixteenth century. The idea had, however, been proposed by the Greek philosopher Aristarchus of Samoa as early as the third century BCE. It was dismissed at the time – which was probably correct since there was then little or no evidence to support it. When, about two hundred years later, Ptolemy devised a detailed model of the solar system that had a stationary Earth at the centre, this was actually the simplest model capable of explaining the available observations. By the time Copernicus came along, more data on the nature of planetary motion was available. He also had the key insight that the orbits of the planets were ellipses, with the sun located at one of the foci, rather than circles with the sun at the centre, as Aristarchus had suggested. Copernicus's resulting theory was able to account for all the known facts. Nevertheless, this heliocentric theory only came to be fully accepted after the invention of the telescope enabled greatly improved measurements of the motion of the planets. Given the context of their theories and the nature of the experimental evidence available, it can be argued that both Ptolemy and Copernicus applied Occam's razor correctly (though not consciously) in reaching their conclusions.

One simplifying assumption, made in all scientific work, is that the fundamental physical laws, such as gravity, are the same everywhere at all times. Thus, although when Alan Shepard played golf on the moon in 1971 he could hit the ball very much further than the best professional golfer on Earth, the same law of gravity applied in both cases. Because the mass of the moon is much less than that of the Earth, the force of gravity on the golf ball is smaller. Similarly, if I measure the time it takes for an object to fall from my hand to the floor today, I will get the same result tomorrow – provided, of course, that I release it from the same position each time. One of the consequences of the assumption that the fundamental laws do not change with time is that magic or miracles, in which the laws of physics are temporarily suspended while some otherwise impossible process is assumed to occur, are not part of science.

Another example of the application of scientific reasoning is the theory of evolution. Before this idea was developed in the nineteenth century, primarily by Charles Darwin, there was no scientific understanding of how living creatures had originated. The word 'creature' is itself a clue to the widely held belief that living beings must have been 'created' by some superior intelligence, generally known as God. That, at least, is the 'creation myth' of the Judeo-Christian monotheistic culture. Darwin's theory has two main strands: first, the realization that living creatures change or evolve gradually from one generation to the next; second, the understanding that although these changes are effectively random, some will result in the offspring becoming better able than their parents to survive the challenges of their environment, while others will have the opposite effect. The offspring inheriting the beneficial changes are more likely to survive and produce further generations so that, over time, the species becomes better adapted to its environment. An increase in the complexity of the organism often results and this explains how the wide variety of living beings on Earth have evolved from very simple chemical molecules over a period of around ten billion years.

Evidence supporting Darwin's theory was found by examining fossils, the preserved forms of living beings that existed in the past. It has also been fully supported by later, twentieth-century, research which identified the chemical processes associated with inheritance and mutation. There are still detailed questions to be answered about how evolution works in various contexts but nothing has been discovered that is inconsistent with the principles of the theory. Darwin's theory is therefore an example of the application of both Popper's falsifiability principle and Occam's razor. An alternative model that requires the postulate of an additional unnecessary 'entity', comprising an all-powerful supernatural being, should, therefore, be rejected. Even if one believes for other reasons that God exists, the superficial simplicity of instantaneous creation has little or none of the explanatory power of Darwin's alternative.

In these illustrations, the application of Occam's razor is quite clear and uncontroversial but this is not always the case. If there had been no planet Mercury or any other evidence for the failure of Newton's theory of gravitation, would Einstein's theory of general relativity ever have been proposed or accepted? One might think not, because Newton's theory appears much simpler and easier to understand than Einstein's, which is mathematically complex and challenging. General relativity, however, is based on very general assumptions about the nature of space and time, while Newton's theory is based on essentially *ad hoc* assumptions about the form of the gravitational force. It can therefore be argued that Einstein's theory contains fewer of the entities that Occam says we should not multiply beyond necessity. Occam's razor is an essential tool to be used in developing scientific theories but it is not always infallible, as there can be controversy as to which of two or more alternative theories is actually the simplest.

Emergence

Fundamentally, the principle of reductionism implies that no new fundamental laws are required to explain higher-level phenomena. New principles often emerge, which are sometimes described as 'laws' (Ohm's law of electrical resistance is an example) but they are always consistent with, and in principle at least derivable from, the laws governing the lower level. The same fundamental physical laws determine both how an atom moves in a vacuum and how it behaves as part of a muscle.

Nevertheless, the question arises as to what, if anything, may be lost in the reductionist process. Does a description of the behaviour of its constituents tell us all that there is to know about an object or a process? Obviously not because, as we go from, say, a biological description of a living object to one in terms of atoms, we lose vitally important aspects of our understanding of the system at every stage.

Consider, as an example, a work of art – specifically, the *Mona Lisa*, painted by Leonardo da Vinci. When viewed from a distance, we see a depiction of a Renaissance woman offering us her famously enigmatic smile. If we examine the picture more closely, we can make out da Vinci's separate brush strokes. If we look at the canvas through a microscope, we see each stroke is composed of small grains of paint of various colours. We might imagine that the reductionist process would entail describing the painting purely in terms of its smallest parts – the grains of paint. Is this all there is to it? Is the *Mona Lisa* just a collection of coloured spots? From one point of view there is nothing else. The grains of paint are the only reality – because, after all, the quality and intensity of colour that an individual grain produces is exactly the same as it would produce in another picture or on the artist's palette. Nevertheless, everyone would surely agree that the painting as a whole possesses properties over and above those of its constituents: the whole is greater than the sum of its parts. This is consistent with the reductionist principle because the higher-level properties (the figure, her smile, and so on) *emerge* from the fact that the lower-level dots have been arranged in a particular way by the artist.

As a further example, consider the page of print you are reading now. It is composed of text, containing sentences made up of words, which in turn are composed of letters of the alphabet. Taking another step down the reductionist road, the letters themselves are made up of dots of black ink on a white background. We could continue down to the atomic level and beyond. Is that a full and adequate description of this text? Surely not, because knowing that the page consists of black ink dots on white paper tells us nothing about the meaning I am trying to convey. The properties of the atoms in the paper and ink are such that they enable the ink to have a black hue and to adhere to the white paper; these would be just the same if the page contained a Shakespeare sonnet, a piece of doggerel, or a meaningless jumble of letters. The higher-level meaning and significance emerge from this lower-level reality.

At the moment, I am typing the text of this book using word processing software that is installed on my personal computer. How might this be described from a reductionist viewpoint? Each time I press a computer key, I send a particular set of electrical impulses from the keyboard, through an interface into the computer's memory, where it is stored in the form of a series of *binary bits*. A binary bit is a piece of electronic circuitry that can be in either of two possible states, which are conventionally called 'one' and 'zero'. In the binary bit representation of a number, the right-hand-most bit represents 1 or 0, the next to the left represents (1 or 0) $\times$ 2, the next one (1 or 0) $\times$ 2 $\times$ 2, and so on. Thus, the number six is represented by 00000110, ten by 00001010, fifty-seven by 000111001 and so on. All the operations of a computer can be described in terms of the manipulation of such bit patterns: for example, the sum $1 + 1 = 2$ is performed by starting from two bit patterns, each corresponding to 1 and creating a new bit pattern equivalent to 2: from 00000001 and 00000001 make 00000010.

The computer operates because a circuit within it represents a bit that allows a current to flow if it has the value 1 but not if it corresponds to 0. When such circuits are connected together, the value of one bit is often controlled by the currents flowing into it from neighbouring bits. This principle can be extended to build complex arrangements of circuits that can carry out mathematical operations. The configuration used in a particular case can also be determined by a *program*, which is essentially a series of bits that are inputted into the computer from an external source such as a keyboard or a disc.

Each letter of the alphabet is associated with a unique pattern of bits. For example, in the ASCII convention the letter 'a' is represented by 001100001 and 'b' by 001100010. Each operation of the keyboard generates a particular bit pattern, which the computer can use to display characters on the screen or print them out.

In this way, the operation of a complex object (the computer) can be reduced to the behaviour of its simpler and smaller components (the binary bits). Is this all that a computer is: a device consisting of a huge number of electrical components each of which can switch between two possible states? This is certainly true at the level of the computer components but there is also a lower-level description (that of the atoms or the fundamental particles) and a higher-level one, such as the text being produced. Once again it depends on what level of description is relevant. At one level, there are really only the bits and the electronics but, if what is written has meaning and significance, this reductionist description of a computer is only a very limited description of what it is and does.

Another word for emergence is *supervenience*: higher-level properties supervene on those of the lower level, so that a work of art supervenes on the pattern of paint grains and a sonnet supervenes on the printed page. An important point to note is that the same emergent properties can supervene on different substructures. A Shakespearean sonnet can be printed using different machines, employing a variety of fonts, while a reproduction of the *Mona Lisa* displayed on a high-resolution display screen possesses many of the same higher-level properties as the original picture. Indeed, although most of us have probably never visited the Louvre in Paris and only seen a reproduction of the painting, we can still appreciate much of its artistic meaning and significance.

In these examples, the emergent properties are the result of human activity. Without Leonardo da Vinci, the *Mona Lisa* would never have existed and without Shakespeare there would be no

Shakespearean sonnets. However, supervenience also occurs when birds build nests from leaves and grass, and insects, such as bees, create complex constructions that are essential to their existence. Figure 1.1 shows an example of a bower-bird's nest: at one level this is a collection of stalks of grass and brightly coloured objects but at another it is an arrangement instinctively created by a male bird to provide a structure that will attract a female as a mate. (The construction is only used for mating – the female bird builds a separate nest in which to lay her eggs.) In the same way as the *Mona Lisa* consists of grains of paint, the bower-bird's nest is 'only' grass and bright objects! Each stalk of grass is subject to the natural laws of botany but its function, which is essential to the preservation of the bower-bird species, supervenes on this.

Figure 1.1 A bower-bird's nest, which is built and decorated by the male bird using grass and bright-coloured objects, such as flowers, berries and plastic objects. A female mate is attracted by the quality of the nest and the richness of the decoration. (Cultura RM / Alamy)

Inanimate objects also demonstrate high-level natural properties: for example, at one level a rainbow is a collection of water drops with sunlight passing through them but at another it displays the familiar pattern of colours, whether or not anyone is looking at it. In all cases, complexity and functionality supervene on the underlying structures without the need for any additional physical laws.

Returning to a more conventionally scientific context, consider the property known as *solidity*. This is a word used to describe any solid object: it implies some degree of hardness and an ability to retain its shape when left alone for some time – millions of years in the case of some rocks. As I shall discuss in later chapters, solids are made up of atoms, often arranged in regular crystalline patterns but the property of solidity cannot be attributed to any one atom. Moreover, although the detailed properties of different solids vary, the general property of solidity is possessed by them all. A rock, a piece of metal and a wooden block are all described as solid, even though their internal compositions differ greatly. Solidity is therefore an emergent property that supervenes on the underlying atoms.

The rest of this book

Having assembled our tool kit, my aim is to use the principles of falsifiability, simplicity, and emergence to go on a journey from the atom, via the human brain, right up to the level of human societies, trying to show how reductionism can provide scientific understanding at every stage.

We shall see that the laws of physics describe quite precisely how fundamental particles join together to form atoms. In simple cases, this enables high-precision calculations of atomic properties whose results can be compared with experiment. The exact agreement that results is extremely powerful evidence for the correctness of reductionism in this context. This is an example of what has been called a *constructive* explanation. In more complicated cases, such as larger atoms, molecules and, indeed, solids, liquids, and gases, such exact calculations are often not practicable. Following our three principles, however, we should assume the validity of the reductionist process unless and until it is falsified.

When we come to biological systems, exact, constructive calculations of the type discussed above are just about impossible. Nevertheless, the chemical structures of important biological molecules, such as DNA and many proteins, are now known and there is no indication that these follow anything other than the basic laws of chemistry. The same is true about the operation of the nervous system and the brain. Once again, reductionism holds, in the sense that the biological components (nerve cells, for example) are subject to the same fundamental laws as everything else. Despite this, the extension of the reductionist principle to account for the emergence of what is called mind or consciousness from the operation of the brain is still quite controversial. Can such properties really supervene on those of the brain when treated as a physical object? If so, where does this leave free will? I shall argue that reductionism is valid even when applied to conscious human beings but I will also consider and discuss the views of some of those who disagree.

The application of reductionism to societies composed of collections of living organisms, such as colonies of insects and flocks of birds, gives considerable insight into the reasons for their behaviour. Using this principle to understand human societies is, however, more difficult and controversial. I shall argue that this is not because the method fails in principle but because the ability of individuals to understand and influence social behaviour greatly increases its complexity.

The one area where the application of the reductionism principle is most in doubt is the strange behaviour that emerges in some experiments relating to quantum physics. The behaviour of large-scale objects, such as the scientific apparatus used to observe and measure quantum systems, seems to follow laws that differ in some ways from those that apply to the quantum objects themselves. I shall discuss this still quite controversial possibility and try to explain why, even if it is correct, it does not seriously challenge the conclusions of the earlier discussions.

2
The building blocks of matter

The word 'atom' comes from the ancient Greeks. The idea probably originated with Leukippos, who was born in Miletus in the first half of the fifth century BCE. It was developed and enhanced by Democritus, one of his pupils, who lived in Thrace from 460 BCE to 370 BCE. These thinkers proposed that if a piece of matter is divided into smaller and smaller pieces, a point is eventually reached where further subdivision is impossible. They called this irreducible minimum *átomos*, which means 'uncuttable' or indivisible and then went on to consider how the properties of these atoms might account for the behaviour of matter. They suggested that a substance like iron is solid and hard because the iron atoms possess hooks which connect them together, while the atoms of liquids, such as water, must be smooth and slippery. Although these ideas have considerable resonance with the modern picture of atoms, this is really only a happenstance: the Greeks had no direct evidence for the existence of atoms or any way of estimating their size or mass. Nevertheless, many of the elements of reductionism are present in their thinking, including the idea that large-scale properties emerge from those of an underlying substructure.

Although the atomic hypothesis attracted interest from later thinkers, including Isaac Newton, the concept was not significantly developed until the early nineteenth century, when John Dalton proposed his atomic theory. Dalton came from a Quaker family, which at that time meant that he was banned from attending any English university. Instead, he obtained a position in

New College in Manchester, which had been founded to provide educational opportunities for those who were not members of the Church of England. Dalton suffered from colour blindness, and one of his earliest scientific papers proposed a theory to account for this condition but today he is remembered chiefly for his atomic theory, which demonstrated how the existence of atoms can account for many of the properties of physical substances.

One of Dalton's key ideas was that chemical compounds, such as water or common salt, are composed of 'molecules', which in turn consist of definite numbers of different types of atom connected tightly to each other. Some substances are composed of atoms of only one type; these are known as *elements*. Notable among the elements are the gases hydrogen and oxygen, as well as metals such as iron and copper. When a compound is formed from the union of two or more elements, the resulting molecules contain a given number of each type of component atom. For example, a water molecule (H_2O) consists of two atoms of hydrogen bound to one atom of oxygen. Dalton performed experiments that confirmed this *law of constant proportions* and he was able to extend these studies to obtain the relative masses of many of the common elements.

It is important to note how the principle of reductionism applies here. If it is valid, no new laws of physics should be required to explain why the properties of liquid water are different from those of hydrogen and oxygen, which are gases. It is the aim of this chapter and the next to explain how this comes about.

Further support for the existence of atoms emerged when other nineteenth-century scientists built on Dalton's work to show that the atomic postulate could provide an explanation for many of the properties of gases. Nevertheless, some others maintained that atoms did not really exist and that the atomic postulate was only an aid to understanding, with no underlying physical reality. The German philosopher, Ernst Mach, who is probably best known for his explanation of what happens when an object moving through a gas breaks the 'sound barrier', was a member of this group. He strongly maintained the view that the purpose of science is to account for the results of experiment and that no reality should be attributed to objects that could not be directly observed – as was then the case for atoms.

The existence of atoms was not generally accepted until the early years of the twentieth century, when Albert Einstein (who was soon to become famous for his theory of relativity) proposed an atomic explanation for the phenomenon of *Brownian motion*. This is named after the early nineteenth-century biologist Robert Brown, who observed how small particles, such as grains of pollen, floating on top of a liquid, move around in an irregular manner. Einstein showed that this could be explained if the grains were actually being jostled by the motion of the atoms or molecules making up the liquid. Whether or not this explanation would have convinced Mach of the reality of atoms is moot, but it led to a general acceptance of the concept by the scientific community. Nowadays, powerful microscopes can form images of individual atoms and any doubts about their reality have been resolved.

What atoms are made of

Until the twentieth century, little was known about the internal structure of atoms and, indeed, they were thought to be fundamental, indivisible objects, as their original Greek name suggests. It is now known, however, that atoms are composed of even more fundamental particles and that their properties, including the capacity to combine together to create molecules, result from this.

To a first approximation, atoms can be envisaged as very small spheres. An atom is a few ten-millionths of a millimetre across, which is around one millionth of the diameter of a typical human hair. All atoms are composed of two kinds of more fundamental objects: a nucleus, where nearly all the mass of the atom resides, and a number of much lighter particles, known as electrons. The nucleus is more than ten thousand times smaller than the atom itself, while the size of the electron is zero – or certainly smaller than can currently be measured. Both the nucleus and the electron carry what is known as an *electric charge*. This is a fundamental property of nature, similar to mass, in that it can only be understood operationally, in terms of its properties. That is, it can be understood only by what it does rather than what it is. The essential properties of electric charge relate to the forces charged objects exert on each other. There are two types of charge: positive (+) and negative (-). Charges with the same sign are subject to a force that pushes them apart, while oppositely charged objects attract. The larger the charges involved, the stronger the forces and the size of the force reduces as the separation between the charges increases. Putting this a little more precisely, the size of the force is proportional to the product of the magnitudes of the two charges and follows what is known as the *inverse square law*; which means that if the separation is doubled, the force reduces by one quarter, while if it is tripled, the force is one ninth of its original value and so on.

An atom consists of a positively charged nucleus along with a number of electrons. Each electron carries the same small quantity of negative charge; the nucleus carries a positive charge that exactly balances the total charge of the surrounding electrons, so that the net charge on the atom is zero. This is a consequence of the fact that every nucleus also has an internal structure and contains some definite number of particles known as *protons*. A proton carries a positive charge whose magnitude is the same as that of the negative charge on an electron. The protons are held tightly together in the nucleus by the *strong nuclear force*, whose details are not relevant to this discussion. The number of protons contained in the nucleus determines to which of the 92 elements the atom corresponds.

The simplest atom is that of the element hydrogen, which consists of a nucleus that is just a single proton, along with a single electron. Like other atoms, hydrogen behaves in many contexts just like a small spherical ball. This immediately raises the question of how this can be consistent with the fact that it is composed of two charged particles of opposite sign. The electron and proton exert forces on each other which tend to pull them together, so why is the electron not drawn right into the nucleus, reducing the atom to a size ten thousand times smaller than it actually is? A possible way to resolve this might be to draw an analogy with the motion of the Earth round the sun, where the centrifugal force tends to push them apart, balancing their gravitational attraction. However, if a similar explanation were applied to hydrogen, the atom would be expected to have the shape of a flat disc rather than a sphere. Moreover, it is known that a rotating charge, such as an electron in orbit around a proton, should radiate energy in the form of light, slow down, and fall on to the proton in a very short time. The resolution of this question requires some knowledge and understanding of what is known as *quantum physics*.

The need for the quantum

Many people nowadays have heard of quantum physics and believe it to be a difficult and mysterious subject. It is certainly true that many of the ideas and concepts underlying quantum physics appear strange at first sight and give rise to conceptual problems that remain controversial to this day. Another potential difficulty lies in the fact that a full understanding of the properties of atoms requires the use of quite advanced mathematics. In discussing atoms, however, these difficulties can be largely avoided, provided we are prepared to accept some of the unfamiliar ideas of quantum physics and take some of the mathematical results on trust.

The first idea I want to discuss is what is known as *wave-particle duality*. Early in the twentieth century it was found that in some situations an object, such as an electron, has properties quite unlike these normally attributed to particles and more akin to those of waves. Anyone who has lived near the sea or has travelled on it will be aware of ocean waves; these can be very large, exerting violent effects on ships and providing entertainment for surfers. For our purposes, it will be more useful to think of the comparatively gentle waves or ripples that result when an object, such as a stone, is dropped into a calm pond. The pattern of ripples on the water surface appears to move outwards from the point where the stone entered the water and because of this, such waves are known as *travelling waves*. Other similar examples are the waves commonly associated with sound and with light. Sound waves consist of small variations of pressure in the air, generated by a source such as a voice or a musical instrument, which travel through the air to our ears, where they cause our eardrums to vibrate and provide the experience of sound. Light waves are vibrations of electric and magnetic fields; unlike sound waves, they do not need a medium for their transmission and can travel through vacuum – as they do when they travel through the space between us and the sun.

Another form of wave, which will be important for our understanding of atoms, is the *standing wave*, where the wave vibrates but does not move through space and is confined within some definite volume. Standing waves are the basis of operation for many musical instruments. In a stringed instrument, such as a violin or guitar, the musical notes are produced by vibrations of the string and these have the form of standing waves, as illustrated in Figure 2.1. The waves shown correspond to three of the possible patterns of vibration of a string with a given length that is held at both ends, so that the displacement of the wave is zero at these points. In a guitar, for example, one end is fixed at the bridge of the instrument and the other at the point where it is

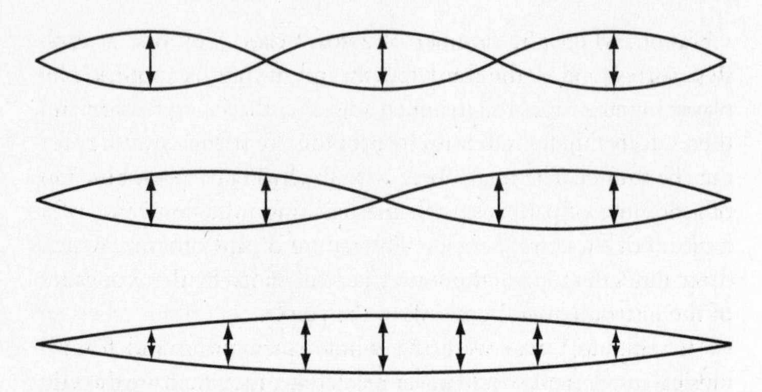

Figure 2.1 Some of the wave forms that can be created on a guitar string. The bottom pattern corresponds to the 'fundamental' with the lowest frequency, while the frequencies of the next two patterns are twice and three times that of the fundamental. (David Berger)

held by the player against one of the frets. When the instrument is played, the string is caused to vibrate and the number of vibrations per second (which may be as low as ten or as large as ten thousand) is known as the *frequency*. The value of this frequency is determined by the tension in the string and the distance between the points where the vibration is zero. If this distance is halved, the frequency is doubled, while increasing the tension also causes the frequency to increase. The distance the string is displaced while vibrating is known as the *amplitude* of the wave.

The pattern corresponding to the lowest frequency, known as the 'fundamental', is the bottom pattern shown in Figure 2.1, where the only points of zero vibration are at the two ends. The other two patterns shown are the 'first harmonic', which has an additional zero point in the middle and therefore twice the frequency of the fundamental, and the 'second harmonic', with two zeros that divide the string into three equal parts; it has three times the frequency of the fundamental. Similar patterns of vibration, where the number of zeros increases by one at each step, correspond to the third, fourth, and further harmonics. The player initially tunes the string by adjusting the string tension and then selects the desired notes by pressing the string down against the chosen fret. The total vibration may actually be a combination of a number of patterns, with the resulting note consisting of a mixture of these frequencies. The nature of this mixture, which determines the tone of the sound, is determined by the properties of the instrument and the skill of the player.

If standing waves were the whole story, the sound from a musical instrument would never reach our ears, which are directly sensitive only to travelling waves in the air. However, when a guitar or violin is played, the body of the instrument oscillates in sympathy with the string and this generates a travelling wave in the air that radiates out to the audience. Much of the science (or art) of designing musical instruments lies in ensuring that the frequencies of the notes determined by the allowed wavelengths of the standing waves are reproduced in the emitted travelling waves. A full understanding of the behaviour of musical instruments and the way they transmit sound to a listener is a major topic in itself, which we do not need to go into any further here.

Waves and particles

For more than a hundred years it was accepted that light was a form of wave motion and this was confirmed by many experiments. In 1865, James Clerk Maxwell, a Scottish scientist and inventor of the first method of colour photography, developed a theory of electricity and magnetism that completely explained the known properties of light in terms of waves. This theory also applies to other electromagnetic radiation, such as X-rays and radio waves.

Towards the end of the nineteenth century, however, experiments were performed which could not be explained by

Maxwell's theory. In these, light was allowed to strike a metal surface, which resulted in electrons being emitted. When the energies of these electrons were measured, they were found to have values that were proportional to the frequency of the incident light. Increasing the intensity (that is, the brightness) of the light increased the number of electrons emitted but if the light frequency stayed the same, the energy of each individual electron was unaffected.

In 1905, Albert Einstein published three papers, each of which has had great significance for our understanding of the physical world. One of these was the paper on Brownian motion mentioned earlier, another was the paper that first set out the theory of relativity and the third provided an explanation for the above phenomenon, which is known as the *photoelectric effect.* Einstein showed that the photoelectric effect could be understood if the energy in the light beam was in the form of packets or *quanta* (singular *quantum*), each of which has an energy whose value is determined only by the frequency of the light wave. The frequency of the light wave and the energy of the corresponding quantum are directly proportional: double the frequency and you double the energy. He also proposed that when light interacts with an electron in the metal, the whole quantum of energy is transferred to the electron. Each electron therefore acquires an energy that is determined by the frequency of the light and independent of its intensity, just as is observed experimentally.

The actual value of the energy of a light quantum is obtained by multiplying the frequency by a universal number known as *Planck's constant.* This is named after Max Planck, another German physicist working in this field at around the same time as Einstein. Planck's constant has a very small value, which largely accounts for it not being discovered earlier. When a process involves many quanta, their individual nature is undetectable and it was not until phenomena such as the photoelectric effect were studied that the quantized nature of light was noticed.

This result, and those from other experiments performed shortly afterwards, implied that in many contexts light has properties similar to those of a stream of particles, which are often referred to as *photons*. This seems quite inconsistent with the picture of light as a wave while a particle model is unable to explain the results of earlier experiments that supported the wave theory. Thus the idea of *wave-particle duality* was born: some experiments appear to demonstrate conclusively that light must be a form of wave, while others can only be explained on the basis that it is a stream of particles. The model used must be chosen to fit the context. Understanding this apparent contradiction has been a challenge since 1905, but it turns out that no actual paradox is implied, because it is not possible to design an experiment where both the wave and particle properties are demonstrated unambiguously at the same time.

The French physicist Prince Louis de Broglie was born in Dieppe in 1893. He was known as 'prince' because his father was the sixth Duc de Broglie and he inherited the title when his brother died in 1960. He worked in radio communications during the First World War and when it was over, focused his attention on the developing field of wave-particle duality. In 1924 he presented his doctoral thesis, in which he speculated that if light can behave as particles, then material objects, such as electrons, might in some circumstances behave like waves. This radical idea was endorsed by Einstein and in 1927 two American physicists, Clinton Davidson and Lester Germer, performed an experiment in which they directed a beam of electrons at a crystal. After the electrons interacted with the crystal, a number of beams emerged whose directions and intensities were essentially the same as those produced by X-rays undergoing a similar process; this had been explained on the basis that X-rays have wave properties. Later experiments demonstrated similar wave properties for heavier particles such as neutrons. It is now believed that all material objects behave like waves in some circumstances.

In principle, even everyday objects such as grains of sand, footballs or motor cars have wave-like features, but they are completely unobservable because the heavier the object is, the less detectable its wave properties become.

A connection between the wave and particle properties of both light and matter can be made by considering what happens when we increase the intensity of a beam of radiation. By 'intensity', we mean the energy carried by the beam, which turns out to be proportional to the amplitude of the wave squared (that is, multiplied by itself). From the point of view of the particle model, this increase in energy corresponds to an increase in the number of particles contained in a given volume of the beam – because each photon carries the same amount of energy. Thus, the density of particles is proportional to the intensity of the wave. This can be generalized to situations where there is only a small number of particles (often only one) under consideration, by postulating that the intensity of the wave at any particular point determines the probability of finding the particle close to that point. The bigger the value of the wave at some point in space, the more likely it is that we will find a particle in the vicinity of that point. One could imagine that this means that the particle is moving rapidly and more or less randomly, except that it spends more time in places where the wave is strong and less time where it is weak, and we shall sometimes use this language in our discussion. The conventional view, however, is not to take this literally but rather to say that if we are not actually observing a particle we should not assume that it even has a position.

In the examples of waves discussed earlier there was always something vibrating: the string moving up and down, the pressure of the gas in a sound wave, the electrical field in a light wave. The corresponding property in the case of matter waves is known as the *wavefunction*. This is not generally directly observable and is best thought of as a mathematical function that allows us to calculate the probability of finding the corresponding particle at a particular place. Another important property associated with the wavefunction is the energy of the particle. Earlier (see Figure 2.1) we saw that the vibrational frequencies of a guitar string have particular values that depend on the shape of the waves excited on them. Also, the energy of a photon is equal to Planck's constant multiplied by the frequency of the associated wave. Similar results hold in the case of matter waves: the form of the wavefunction associated with a particle such as an electron is directly related to the value of the electron's energy (although the calculation is less straightforward than in the case of either guitar strings or photons).

Back to the hydrogen atom

I want to build on the above ideas to show that the form of the matter wave representing an electron in a hydrogen atom can only take one of a set of particular patterns and therefore that the value of the atom's energy is restricted to a set of particular values. I shall again use the analogy of the vibrating guitar string, although this will get us only so far in the case of the atom.

As discussed earlier, the hydrogen atom contains a single negatively charged electron that is attracted to the positively charged nucleus at its centre by an electrical force but there is something that prevents the nucleus and electron from collapsing into each other. To understand how this can be explained using quantum physics, we must first note that the wave properties of the heavy nucleus are largely negligible while those of the electron play an important role. What form might we expect the wavefunction associated with the electron to have in this case? Because the electron is pulled towards the nucleus by the electrical force, even though it does not collapse into the nucleus, it will be found somewhere in its vicinity. This means that the probability of finding the electron an appreciable distance away from the nucleus must be very small, implying that the wavefunction is effectively zero in this region. In the earlier discussion of the waves on a string (see Figure 2.1), we saw that the condition that the vibration must be zero at both ends means that only specific patterns of wave are allowed, each of which corresponds to a musical note with a particular frequency. In a similar way, the requirement for the wavefunction associated with the electron in a hydrogen atom to be zero some distance from the nucleus helps determine the form and shape of the allowed patterns, which now correspond to particular values of the energy.

Continuing with this analogy, we can expect the wavefunction corresponding to the lowest energy state of the atom to be similar to that associated with the fundamental note of the guitar string. That is, it should be zero far from the atom on both sides and build up to a peak at the centre. One important difference between a guitar string and an atom, however, is that the string is a one-dimensional object – it exists only along the line of the string – while an atom is three-dimensional. The force between the electron and the nucleus always acts along the line connecting these two objects, which means that it will be the same whatever the direction of this line in space. Putting these two ideas together suggests that the wavefunction should resemble a spherical cloud surrounding the nucleus. This is illustrated in Figure 2.2(a).

This argument explains how the wave model of the electron resolves the problems raised earlier when we considered how the hydrogen atom might form. First, because the wave pattern is spherical, the probability of finding the electron at a point depends only on the distance of this point from the nucleus and not on its direction: in this sense the electron really does 'surround' the nucleus. Second, if the electron were to collapse into the nucleus, there would be a large probability of finding it there and a zero probability of finding it anywhere else, and a wavefunction with these features does not correspond to one of the allowed patterns. Referring once again to the guitar-string analogy, there is no allowed vibration where the amplitude is zero everywhere along the string apart from a large peak at the centre.

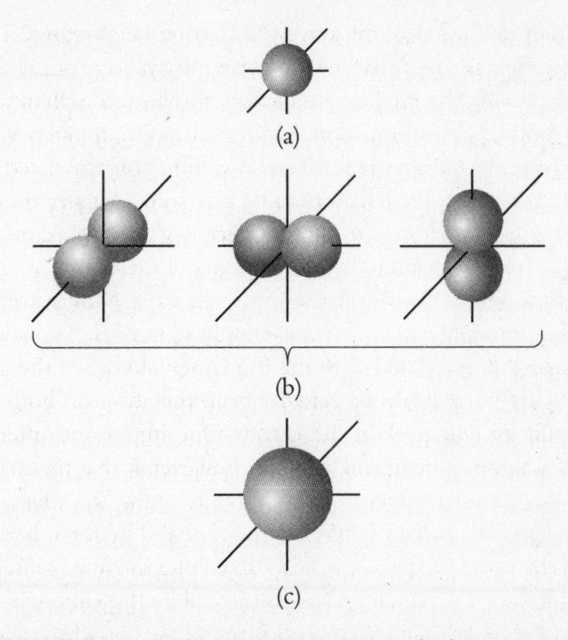

Figure 2.2 The wave patterns corresponding to the lowest two energy states in hydrogen. (a) shows the pattern corresponding to the lowest energy (the ground state); (b) and (c) respectively show the three dumb-bell shapes and the spherically-symmetric pattern, all of which correspond to first-excited states. (David Berger)

In the same way as the lowest energy state of hydrogen (which is often termed the *ground state*) is analogous to the lowest frequency vibration of the guitar string, higher energy states (often known as *excited states*) can be compared with the other allowed patterns of string vibration. The first harmonic illustrated in Figure 2.1 has two peaks and a zero value at the centre. Using this, we might expect that the wavefunction corresponding to one of the excited states of the hydrogen atom should have a zero value at the nucleus, increase with distance away from the nucleus and then decrease to become zero again at large distances. This is illustrated by the three shapes shown in Figure 2.2(b), which resemble figures-of-eight or dumbbells centred on the nucleus. Unlike the ground state, these wavefunctions are not spherical, which appears to be inconsistent with the argument set out above. In fact, the three dumbbells point in mutually perpendicular directions and all three states have the same energy. This means that it is not possible to know which state is occupied by the electron. The probability of finding it at a particular point in space is therefore determined by the average of the three shapes, which turns out to depend only on its distance from the nucleus and not on its direction.

As well as the three dumbbell shapes, there is one more, with very similar but slightly lower energy. Its wavefunction is spherical, as shown in Figure 2.2(c). All four are termed 'first excited' states. The next set (the 'second excited' states) consists of four that are similar to the first excited states, along with a further five whose shapes are even more complex. At even higher energies, further sets containing mixtures of more complex patterns emerge.

This is about as far as it is possible to go using this type of argument. Further progress involves the use of mathematical equations and any attempt to do this in detail would certainly not be appropriate for this book. A general description of the process may, however, be a help in understanding what is involved. To make a detailed study of any kind of wave motion, physicists first derive a mathematical equation whose solutions describe the form of the wave. In the case of a guitar string, such an equation can be derived that connects the motion of every point on the string to the tension in the string and its density (mass per unit length in this case). It turns out that the only allowed solutions to this equation correspond to the string vibrating at particular frequencies and with the patterns of vibration shown in Figure 2.1.

In the quantum case, the analogous equation governing the behaviour of matter waves is known as the *Schrödinger equation.* Erwin Schrödinger was born in Vienna in 1887. His academic career involved an unusually large number of changes of residence. He worked in Vienna, Wroclaw in Poland, Zürich, Berlin, Oxford, Vienna again, and Dublin. Some of these moves were not made for scientific reasons: he was strongly opposed to Nazism and could not live in Germany after 1933 or in Austria after the *Anschluss*. He also had an unconventional lifestyle, in that he lived with two women in a *ménage à trois*, which was not acceptable to the establishments of some institutions at that time.

In 1926, while a professor of theoretical physics in Zürich, Schrödinger published a paper proposing a particular equation whose solutions determine the allowed values of the energy of a given physical system, along with the corresponding forms of the wavefunction. This is the Schrödinger equation; one of its early successes was its application to the case of an electron and nucleus interacting, as in the hydrogen atom. The form of the electrical interaction between the proton and the nucleus is included in the equation and its solutions determine the energies and the corresponding wave patterns, including those illustrated in Figure 2.2.

Strong confirmation of the correctness of this calculation is obtained when its predictions for the energy of a hydrogen atom are compared with those measured experimentally. At ordinary temperatures, an atom of hydrogen is normally in its ground state but it can be excited into one of the higher energy states in a number of ways – for example, by passing an electric current through a container of hydrogen gas. A very short time after being excited, the atom returns to the ground state, losing its extra energy by emitting a quantum of light (that is, a photon). The frequency of the light is related to the energy of the photon so the light emitted in this way has a definite frequency. When this is measured, and the value multiplied by Planck's constant, the result is found to correspond exactly with the value calculated from the difference in the energies of the ground and excited states using the Schrödinger equation.

An illustration of the fact that atoms exist in particular energy states and a particular frequency of light is emitted when the atom makes a transition between two states is demonstrated by sodium lamps that are used in some forms of street lighting. The strong yellow colour of the emitted light is associated with a particular frequency and therefore with the energy difference between two of the quantum states of the sodium atom.

Many more comparisons between theory and experiment have been made in the ninety or so years since Schrödinger first proposed his equation. It has been refined in a number of ways (for example to include the effects of relativity) but nothing has been found that would falsify the principles of his theory or of quantum physics in general.

Beyond hydrogen

Atoms other than hydrogen contain more electrons and have a correspondingly larger positive charge on their nuclei. The next up from hydrogen is helium, which consists of a doubly charged nucleus surrounded by two electrons. This means that the attractive force between the nucleus and an electron is twice as large as in the case of hydrogen, although this is approximately balanced by the repulsive force between the electrons. Both these factors have to be included when calculating the allowed wavefunctions and energy levels of helium using the Schrödinger equation, which appreciably complicates the process. The net result, however, is that the shapes of the wavefunctions are quite similar to those shown in Figure 2.2. In particular, the ground state of helium consists of two electrons sharing a spherically symmetric state similar to that associated with the ground state of hydrogen.

Other atoms have more electrons, so their ground states might be expected to consist of all their electrons occupying a state similar to the lowest energy state found in hydrogen and helium. This is not the case, however, because the properties of atoms with more than two electrons are subject to a condition first proposed by Wolfgang Pauli. Like Schrödinger, Pauli was born in Vienna and he was a professor in Zurich in 1924 when he proposed that there must be a rule requiring that no more than two electrons can be associated with any single quantum state: any others must be excluded, so this rule is now known as the *Pauli exclusion principle*.[1]

Before returning to the question of atoms with more than two electrons, let us first consider the helium atom in a little more detail. We have seen that its ground state corresponds to both electrons being in their lowest energy state. If a third electron were to be added, the Pauli exclusion principle means that it would have to go into a state of higher energy. There would then be three negatively charged electrons repelling each other and only a doubly charged nucleus holding them together. The net result is that the extra electron is actually repelled by the rest of the atom. Helium is therefore very stable, with its two electrons tightly bound to the nucleus and any extra electrons that come close to the atom are repelled. A further consequence of this is that helium atoms are only very weakly attracted to each other, which means that helium is a gas, not only at room temperature but right down to 4.2K – that is, four point two degrees above the absolute zero of temperature (about –268°C).

The element lithium has a triply charged nucleus and three electrons; two of these occupy a state similar to the ground state shown in Figure 2.2(a), although a little smaller, because the triply

[1] The Pauli exclusion principle actually states that only one electron can occupy any quantum state. However, electrons have a further property, known as 'spin'; any given electron can have one of only two possible values of spin. Thus two (and no more than two) electrons can be accommodated in a given state, provided they have different values of spin. Spin gives rise to some other properties, which are not discussed in this book.

charged nucleus pulls the electrons closer towards it. Because of the exclusion principle, however, the lowest available state for the third electron is one with a wavefunction similar to that corresponding to the spherically symmetric excited state of hydrogen (Figure 2.2(c)). The energy of this state is only a little below that of an electron that is completely detached from the nucleus. As a result, when a large number of lithium atoms come together to form a solid, the outer electrons are no longer associated with particular lithium atoms but are free to move through the whole solid. Lithium is therefore an example of a metal, because when an electric voltage is applied to it, the 'free' electrons move, creating an electric current. All common metals, such as copper or iron, contain free electrons that behave in a similar way.

The next atom on is beryllium, with four electrons: two in each of the spherical states. When we come to boron, with five electrons, one of these must occupy one of the three dumbbell states. Further electrons are added to these states as we proceed through carbon (six electrons), nitrogen (seven), oxygen (eight), fluorine (nine) and neon, whose ten electrons fill all the states illustrated in Figure 2.2. Any further electrons have to be accommodated in even higher energy states, so neon, like helium, is particularly stable and unreactive. For historical reasons, helium, neon and other similar elements are known as the *rare* or *noble gases*. The next element is sodium (11), which, like lithium, has one electron outside a core formed by the nucleus and the inner electrons. Its physical and chemical properties are quite similar to those of lithium and this similarity persists as more electrons are added. For example, magnesium (12 electrons) is similar to beryllium (4) while silicon (14) has properties in common with carbon (6) and so on. Further repeats occur as we continue to add electrons, something that was first noted by the Russian physicist, Dmitri Mendeleev, who constructed a 'periodic table' of the elements in the nineteenth century – long before quantum physics was developed.

In elements with larger numbers of electrons, the increased charge on the nucleus results in a further reduction of the size of the low energy states, while the sizes of the most excited states are comparatively unaffected. Overall, this balance between the attractive pull of the nucleus and the repulsion between the electrons ensures that there is no great variation in the sizes of different atoms. For example, the size of the uranium atom, which with ninety-two electrons is the heaviest naturally occurring element, is only about four times that of hydrogen.

The principles of reductionism are illustrated in the properties of the atoms. An atom can be *reduced* to a positively charged nucleus and some negatively charged electrons, subject to the laws of electrical attraction and repulsion and the principles of quantum physics. New properties such as the atom's size and shape *emerge* in this process.

Atoms join together

We began with the idea that atoms are the fundamental building blocks of matter. Material bodies are made up of huge numbers of atoms: a tiny grain of sand contains about ten billion (10^{10}) of them. This implies that matter, at least when in the form of a solid, consists of atoms that are connected to each other. These connections are made in a number of different ways, depending on the substance in question. When two or more atoms join together, the result is called a *molecule*. How and why molecules form is therefore an essential step towards understanding how the properties of matter emerge from the application of quantum physics to their fundamental constituents.

As we have seen, the simplest atom is that of hydrogen, which consists of a single electron attached to a proton. It turns out that the simplest molecule is formed when two hydrogen atoms join together. To understand this, first recall that the lowest energy state of the hydrogen atom is one where the wavefunction is shaped like a sphere centred on the nucleus, which consists of a single proton. Now suppose that two protons are somehow held a short distance apart and that two electrons are added to this set-up, one at a time. The first electron is attracted towards both positively charged nuclei and should therefore be found in the vicinity of one or other of them with equal probability. The resulting wavefunction (remember that its intensity is a measure of the probability of finding an electron at a point in space) should therefore be oval in shape, with its long axis lying along the line joining the two protons. If a second electron is added, it should share the same state as the first, in the same way as the two electrons in helium occupy a state that is similar to the ground state of hydrogen. This picture is confirmed when detailed calculations are performed using the Schrödinger equation.

Now consider how the energy of this set-up might compare with that of two widely separated hydrogen atoms. First, because each electron is attracted to both nuclei, the total attraction should be greater than in the atomic case and should increase the closer the nuclei are to each other. Second, there is a mutual repulsion between the electrons, which is not there when the atoms are far apart. This is difficult to calculate accurately because it depends on where each electron is at any time but on average it will be weaker than the attraction, because each electron is attracted to both nuclei but repelled by only one other electron. Finally, there is a repulsive electric force between the two protons, due to their positive charge; this increases as the nuclei come closer together. There should therefore be a particular separation, where these three forces cancel out and the energy is at a minimum. These conclusions are confirmed by a detailed calculation based on the Schrödinger equation, in which we find that the ground state energy of the electrons is less than that of two separated hydrogen atoms and that this energy initially reduces as the protons approach each other and increases again when they come together. At a particular separation the energy is at its

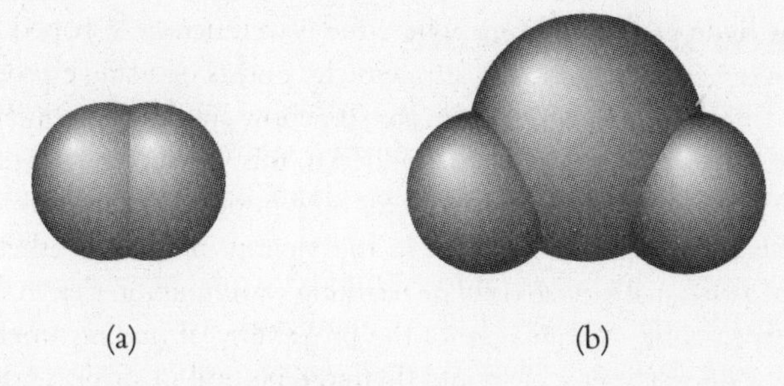

Figure 2.3 The electron wave patterns associated with the molecules of (a) hydrogen and (b) water. (David Berger)

minimum value, so a stable molecule is formed, as predicted above. The distance between the protons in this case equals 0.74×10^{-10} metres and this value agrees with the results of experimental measurements. The shape of the electrons' wavefunction in this case is shown in Figure 2.3(a).

In summary, the hydrogen molecule consists of two hydrogen atoms sharing both their electrons so that they are bound closely together by what is known as a *chemical bond*. Chemical bonds are a very important feature of chemistry and play a vital role in the construction of the molecules that compose the materials that make up much of our everyday material world.

In many ways, therefore, the hydrogen molecule is rather like the helium atom: both consist of two electrons interacting with two positive charges (although the latter are combined in one nucleus in helium). Like helium, the hydrogen molecule is very stable and a third electron cannot be added to it. Also, the interaction between hydrogen molecules is very weak, though not quite so weak as in helium, because the hydrogen molecule does not have the spherical shape possessed by the helium atom, which results in a small residual electrical interaction between pairs of hydrogen molecules. As a result, hydrogen liquefies at a temperature of about 20K (about –253°C) compared with 4.2K for helium.

The water molecule

The bonding mechanism in the hydrogen molecule, whereby two electrons are shared between two atoms, is widely used in nature and forms the basis of many chemical bonds. A common example is the water molecule, which, as is very well known, contains two hydrogen atoms bound to one of oxygen; it is denoted by the symbol H_2O. As mentioned earlier, the oxygen atom consists of a nucleus carrying eight positive charges, surrounded by eight electrons. Two of these electrons are tightly bound to the nucleus in the lowest energy state, a further two occupy the excited state with spherical symmetry and the remaining four are associated with the three dumbbell-shaped wavefunctions shown in Figure 2.2(b). Suppose that one of these dumbbell states contains a single electron and that a hydrogen atom (also carrying a single electron) is brought up towards it. Just as in the hydrogen molecule, the energy associated with these two electrons will be lowered if there is a significant probability of finding both of them in the vicinity of both nuclei. Remembering that this probability is determined by the intensity of the wave, the greatest stability will be achieved if the hydrogen atom approaches along the axis of the dumbbell, in which case a bond is formed when the proton is at an optimum distance from the oxygen nucleus. When a second hydrogen atom is added, similar considerations apply and it will seek to form a bond along the axis of one of the other dumbbells. As a result, the three nuclei might be expected to form a right-angled triangle, with the oxygen at its apex. However there is one more factor to be taken into account: because the two hydrogen nuclei are now quite close to each other, there is a significant electrical repulsion between them, which acts to increase this angle. When this is taken into account, we find that the angle between the bonds is increased from 90° to about 105°. The resulting shape of the water molecule formed in this way is shown in Figure 2.3(b).

The remaining two electrons associated with the oxygen atom occupy the third dumbbell orbital, forming what is known as a *lone pair*. Because all the electrons are now paired up, the exclusion principle means that no more atoms can be attached and the water molecule is stable. The way the electrons are distributed leads to a further property that plays a very important role in determining the bulk properties of water and ice. Because the electrons associated with the hydrogens spend some of their time near the oxygens, their average position is pulled away from the hydrogen nucleus towards that of the oxygen; this means that, on average, the negative electron charge in the vicinity of each hydrogen atom does not quite cancel the positive charges on the hydrogen nuclei, while that on the oxygen atom is a little greater than the charge on the oxygen nucleus. The net effect is that each hydrogen atom carries a small positive charge and the oxygen atom has a net negative charge located on the side opposite to where the hydrogens are attached. This property plays an important role in determining the properties of bulk water and ice, as discussed in the next chapter.

These two examples illustrate how the principle of reductionism applies to the relationship between chemistry (which is the science of molecules) and the physics of atoms. The laws of quantum physics mean that negatively charged electrons can interact with positively charged nuclei to form atoms of particular size and shape. When different atoms come together, they can share electrons to form molecules with shapes and sizes that are determined by the same quantum laws that applied to the atoms. New properties emerge as a result of this process. Thus, when two electrons combine with a doubly charged nucleus to form an atom of helium, a spherical object is created; while two hydrogen atoms can form a molecule with an oval shape and a definite size. When two hydrogen atoms come together with oxygen to form a water molecule, a triangular object, with a definite size and shape, is formed. The concepts of bond length (the distance between adjacent atoms in a molecule) and bond angle (the angle between neighbouring bonds) are not possessed by the individual atoms before the water molecule is formed and are certainly not properties possessed by the three nuclei and ten electrons that constitute the molecule before they interact. This is a further example of how higher-level properties emerge from lower-level constituents without the need to introduce any new physical laws to act at the higher level. Only the fundamental properties of electrons and nuclei (such as mass and electric charge), together with the universal laws of quantum physics, are required to produce all chemical molecules.

One caveat: the calculations involved in solving the equations of quantum physics can be carried out exactly only in some simple cases. It is probably safe to say that the quantum physics of the hydrogen atom is completely understood but the calculation of the detailed properties of other atoms requires the use of approximate methods and sophisticated numerical calculations. In the case of helium, these produce very precise results that are in full agreement with experimental measurements but this precision decreases as the complexity of the atoms and molecules increases. Nevertheless, the accuracy of such calculations has improved immensely as the power of computers has increased and every time some property of a more complex molecule is evaluated, it is found to be in agreement with the corresponding value measured experimentally. Certainly, nothing has emerged from this process that would imply falsification of the basic laws of physics or of the validity of applying the reductive process at this level.

Inside the nucleus

In addition to protons, most nuclei also contain a number of uncharged particles, known as *neutrons*, which are neither attracted nor repelled by the negative electrons or the positive protons.

The protons and neutrons are in turn composed of three more fundamental constituents, known as *quarks*; each proton or neutron contains three quarks which are bound together by the *strong nuclear force*.

This force is also responsible for the strong connections between the protons and neutrons in the nucleus. I do not, however, intend to discuss this any further but rather to take the nuclei as given and build upwards from the atom. This may seem an arbitrary choice but there are good reasons for it. First, to go below the level of the nucleus would require a more sophisticated understanding of quantum physics and quantum field theory than is appropriate in this book. Second, once a nucleus has been formed by combining protons and neutrons, it generally behaves as a very stable object whose properties, such as its charge and mass, are unaffected by anything we do to the atom. This is because the forces that bind the nuclear particles together are very much stronger than those binding the electrons to the nucleus, so the latter maintains its stable configuration in most contexts, including the temperatures commonly encountered on the Earth's surface. Only when temperatures have the very high values typical of the interior of stars like our sun do some nuclei break up and react, releasing the energy that allows the sun's temperature to be maintained. The most important exception to this general rule arises when some very large atoms, such as uranium, display radioactivity. In this case, unstable nuclei can lose some of their constituent particles or break up in the process of fission, the fundamental process exploited in nuclear reactors and power stations. I shall not discuss these processes further except to note that they are well understood and are direct consequences of the fundamental laws of physics when applied to the nucleus.

3

Why water is wet and ice is hard

> The whole surface of the ice was a chaos of movement. It looked like an enormous jigsaw puzzle, the pieces stretching away to infinity and being shoved and crunched together by some invisible but irresistible force. The impression of its titanic power was heightened by the unhurried deliberateness of the motion. Wherever two thick floes came together, their edges butted and ground against one another for a time. Then, when neither of them showed signs of yielding, they rose, slowly and often quiveringly, driven by the implacable power behind them.

This passage, from Alfred Lansing's book *Endurance*, describes the state of the Weddell Sea, a large bay in the north coast of the Antarctic continent, at the time when Ernest Shackleton's ship, *Endurance*, was trapped there in 1914. Shackleton was one of the explorers who attempted to open up the Antarctic continent in the early years of the twentieth century. The Anglo-Irish Shackleton was a member of several Antarctic expeditions before others reached the South Pole in 1912. Subsequently, he decided to mount an expedition to cross the Antarctic continent 'from sea to sea', setting out in *Endurance* in 1914. However, this project came to a premature end when his ship was trapped in the Weddell Sea, where it was eventually destroyed by the forces of the ice and abandoned. The 800-mile journey in an open boat that eventually led to the rescue of the whole crew is often cited as an outstanding example of heroism and endurance.

The reason *Endurance* was trapped and eventually destroyed was because a drop in temperature had resulted in the formation of ice floes on the surface of the water after the ship had entered the sea. These floes were driven further into the bay by strong northerly winds; they collided and jostled with each other for a share of the confined space in the bay and put the ship under pressure which steadily increased to the point where the hull was crushed and broken.

Both the ice floes and the water they floated in were composed of identical molecules; each consisting of an oxygen atom bound to two hydrogen atoms. The hardness of ice and the wetness of water are properties that emerge when large numbers of these molecules are brought together; which of these properties is manifest depends mainly on their temperature. If this process were not so familiar, it might have been thought that the very different properties possessed by ice and water implied that these are fundamentally different substances with different basic constituents.

Bulk matter acquires properties over and above those associated with its component molecules, and these bulk properties can be radically different depending on the environmental context. How does this come about? Is it because the fundamental laws applying to water and ice are different from those governing the behaviour of the component individual atoms and molecules? If so, this would falsify the reductionist principle but if not, the properties of solids and liquids must supervene on the same laws that govern the behaviour of the component molecules. The aim of this chapter is to show how this comes about.

The first step is to acquire some understanding of what is meant by temperature. To begin the process I want to look again at the concept of energy, which was discussed earlier in the context of atoms. An atom prefers to be in the lowest energy state; if raised to a higher energy state, it undergoes a transition back to the ground state, emitting the extra energy in the form of radiation. This process illustrates two important properties of energy. The first is that energy is conserved, which means that it cannot be created or destroyed but only transformed from one form to another. The amount of energy carried away by the emitted radiation therefore exactly equals the difference between the energies of two states. The second property is the idea of *preference*: why does an atom spontaneously decay from a higher energy state to the ground state, emitting radiation, while the reverse process occurs only if an experiment is carefully arranged so that sufficient radiation of the appropriate wavelength is directed at the atom? The simple answer is that the emitted radiation goes out in all directions more or less at random, while the chances of it returning in a focused form that can re-excite the atom are very small – unless the situation is deliberately contrived to achieve this. This tendency for physical systems to change so as to increase their overall randomness plays an important role in many branches of science.

To understand this further, imagine dropping a tennis ball on to a hard surface from a height. Because of the force of gravity, the ball falls towards the floor, which results in a reduction of the gravitational energy associated with the interaction between the ball and the Earth. Where does this energy go? The answer is into *kinetic energy*, the energy associated with the ball's motion, which depends only on the speed of the moving object and not on its direction of motion. When the ball hits the floor, it still has this kinetic energy and, unless it can lose it in some other way, the only thing it can do is to bounce upwards at the same speed it had when it landed. As it moves upwards, the gravitational force pulls it back so that its speed becomes less, its kinetic energy reduces and its gravitational energy increases. Eventually, it pauses instantaneously at the same height as it started from and then starts to fall down again.

If this were all there were to it, the whole process would go on for ever but in practice, the height of each bounce is less than that of the one before and sooner or later the ball comes to rest on the ground. This means that if the total energy is to be conserved, some of it must be converted into another form (that is, not kinetic or gravitational energy). This mainly takes place when the ball strikes the floor, at which point some energy is lost from the bouncing ball to the surface it strikes, which is heated to a higher temperature as a result. In some cases, this energy loss is so large that it absorbs all the kinetic energy and no bounce occurs: for example, if a ball is dropped on to a soft surface, like a cushion. This energy differs from kinetic and gravitational energy in that while energy can be easily transformed from kinetic to gravitational and back again, the process where energy dissipated in the surroundings changes back into kinetic energy does not normally occur. If we simply warm up a surface and place a ball on it, we do not expect it to gain kinetic energy and levitate spontaneously!

This leads to a fundamental principle, which is that there are two kinds of energy transformation: 'reversible' changes, where energy moves from one form to another and back again (as when the ball is bouncing), and 'irreversible' changes, where the process occurs in one direction only (as when it loses energy to its surroundings). The key reason is that in the latter case, the energy is contained in the random motion of the component atoms and molecules, so the chance of them combining to push the ball upwards is extremely small. Randomness operates in this case in much the same way as it does in the case of the radiating atom.

Atoms in a gas

These ideas can be developed further by considering the properties of a gas. The simplest example is one formed from separate spherical atoms (like helium or one of the other rare gases referred to in the previous chapter) contained in some form of closed vessel. Let's assume that atoms are like small hard balls that bounce off each other when they come into contact. Before and after such a collision, the atoms have kinetic energy associated with their motion; assuming that no other form of energy (such as radiation) is involved in the process, the principle of energy conservation ensures that the total kinetic energy of the pair is the same after the collision as it was before. Similarly, when an atom collides with one of the boundaries of the containing vessel, it will bounce off, and there is no loss of energy in this process either. This means that the total kinetic energy of all the atoms in the gas remains the same, even though the way it is shared between the individual atoms keeps changing.

Treating the atoms of a gas as equivalent to hard balls can be justified by considering their quantum properties as discussed in Chapter 2. Because a fixed amount of energy is required to excite an atom from its ground state, this will not occur if the energies of the colliding atoms are smaller than this amount – which is always the case in the situations to be considered. The atoms can therefore be assumed to remain in their ground states at all times and their total kinetic energy before and after a collision must remain the same.

A typical container of gas contains a huge number of atoms, so it is a practical impossibility to work out what happens to each of them individually as collisions proceed. It is, however, possible to analyse their overall behaviour statistically, so as to answer questions such as 'What is the average position of an atom in a gas, how fast is it moving and in which direction?' Everyday experience tells us that if we sit still in a closed room, the concentration of the air is the same in the middle as it is near one of the walls and is the same near the floor as near the ceiling. That is, atoms are expected to be evenly distributed throughout the volume of the containing vessel. Our experience also leads us to expect the air in a closed room to be 'still' – in the sense that there are no winds or draughts. This means that, although all the air molecules are moving rapidly, on average they are not moving in any particular direction. If this is not the case, it is because the air is being disturbed, for example by a fan. If so, this motion stops very soon after the fan has been switched off, showing that the atoms quickly revert to a state in which all directions of motion are equally likely.

Although these experiences are familiar, why do atoms behave in this way? How does the fact that the atoms in a gas are uniformly distributed in space and have no preferred direction of motion follow from the same laws of physics that apply at the level of individual atoms and molecules? To see how this comes about, imagine that the container is empty and divided into two equal halves, labelled left (L) and right (R). Now imagine that three atoms (labelled A, B, and C) are placed in the container – they may all be on one side or the other, or shared between the two sides. There are therefore eight different possible arrangements, as shown Table 3.1.

In two of these arrangements, all three atoms are at one side (left or right), while in the other six, two are on one side and one on the other. If the atoms are inserted at random, it is therefore three times more likely that two will be on one side and one on the other than that all three will be on the same side. If we follow the same procedure with four atoms, there are sixteen possible

Table 3.1

L	*R*
A B C	
A B	C
A C	B
B C	A
A	B C
B	A C
C	A B
	A B C

arrangements: in six of these there are two atoms on each side, in four cases three on the left and one on the right (and another four with their roles reversed), while there are only two configurations where all the atoms are on one side or the other.

The more atoms that we put in the container, the more likely they are to be evenly distributed between the two halves. Figure 3.1 shows the number of possible arrangements of ten atoms. The total number of arrangements is 1024 and of these, only two have all ten atoms on one side, while in more than two-thirds of the cases the atoms are distributed either equally or with six on one side and four on the other. This random distribution of atoms between the two sides is similar to that resulting from the tossing of coins: the chances of a single coin coming down head or tail are 50:50 but if a large number of coins are tossed, they are very likely to come down with equal, or nearly equal, percentages of heads and tails. As the number of coins is increased,

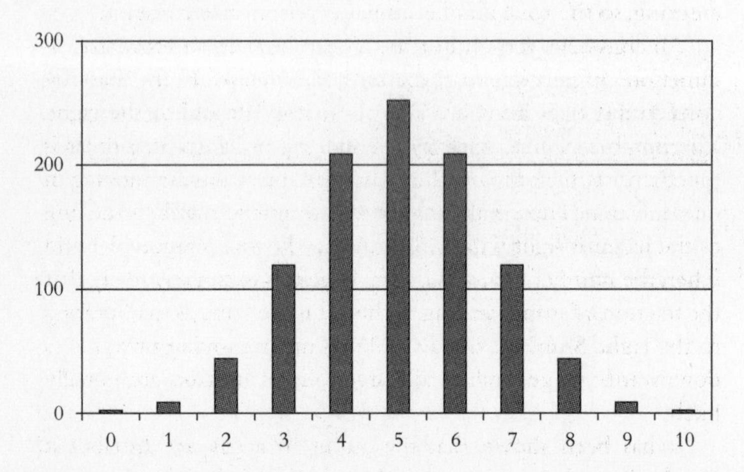

Figure 3.1 The vertical lines represent the number of ways of arranging ten atoms between the two halves of a container. The number at the bottom of each line equals the number of atoms in the left half. (David Berger)

it becomes more and more likely that close to half will come down 'heads' and the rest 'tails'.

The same principles apply if we imagine our container to be divided into more than two parts. Provided the partitions are all the same size and that each is large enough to contain a reasonably large number of atoms, it is extremely likely they will all contain close to the same number. In air, the number of atoms is so large that a tiny cube with sides one thousandth of a millimetre long contains about a million atoms, so a distribution where all such cubes contain close to the same number of atoms is by far the most likely arrangement. The principle that a system containing many particles will adopt its most probable configuration is an essential assumption underlying all such processes. In the case of a gas, it occurs because the atoms frequently collide with each other and with the walls of the container. Any given volume will have atoms continually entering and leaving it; on average, the number leaving is expected to be very close to the number entering, so the total number remains very nearly constant.

An argument very similar to this can be used to show that all directions of motion of the atoms are equally likely. Start by considering how many are moving to the left and to the right. The number of arrangements where these numbers are similar is much greater than the number where all the atoms are moving in one direction. Thus, in the case of ten atoms, the numbers moving to the left and right will be the same as shown in Figure 3.1 and when the number of atoms is very large it is extremely likely that the fraction of atoms moving to the left is the same as that moving to the right. Similarly, there will be as many moving upwards as downwards; to generalize, all directions of motion are equally likely.

It has been shown that the atoms in a gas are distributed evenly throughout the containing vessel and that their motion is evenly distributed among the possible directions. However, there remains the question of how many atoms are moving at any particular speed or, equivalently, how the total kinetic energy is distributed among them. In principle, if all we know is the total energy, this could be shared out among the atoms in a large number of different ways. For example:

- all the energy might be concentrated in a single atom moving at a very high speed, while the others are standing still
- all the atoms might be moving with the same speed
- some atoms could be moving slowly, while others are moving faster and yet others faster again.

You should pause at this point and decide which of these motions you might expect to be most likely. I expect it will take very little thought to reject the first one: if this were to be the case at some point in time, collisions with other atoms would quickly ensure that the energy was shared out among them. Also, if an atom were to be at rest, it would not be long before another one came along, collided with it and set it in motion. On the other hand, if at any time all the speeds were the same, a collision between two atoms is likely to result in one moving faster and the other slower than before. Putting all this together, it seems likely that the atomic speeds are not all the same but are probably not very different from that corresponding to the average energy. This argument can be extended mathematically using statistical principles to calculate what fraction of the atoms may be expected to be moving at a given speed. We find that the majority of atoms have speeds between half and twice the value corresponding to the average energy, while a minority of the atoms are moving much faster than this and a very few are almost stationary. Which atoms move at a particular speed changes with time, as the atoms collide with each other and the sides of the container but the average number stays the same. Given that the number of atoms is very large, this distribution of speeds is by far the most probable arrangement.

Measuring disorder

If, for some reason, the atoms in a gas do not have the properties described above, they very quickly acquire them. Suppose all the atoms are confined to one side of the container by a partition which is then removed. Some of the atoms will very quickly move into the empty side and this will continue until they are evenly distributed. Similarly, if the initial distribution of speeds is not that discussed above, it quickly becomes so as collisions between atoms take place. This is because, as has been emphasized, the state of the system after a large number of collisions have occurred is the most probable state – in the sense that it is the one with the largest number of possible arrangements of the positions and velocities, consistent with the given value of the total energy.

Another way of expressing this is to say that this arrangement is one in which the *disorder* of the system has a maximum value. We can define the number of possible arrangements of the atomic positions and velocities as a measure of this disorder. In the example illustrated in Table 3.1, this number is six for the configuration where two atoms are on one side and one on the other and only two for that where all the atoms are on one side. In fact, the quantity used by scientists is not this actual number but its logarithm, which is called the *entropy*. Non-mathematical readers who do not know, or have forgotten, what a logarithm is should not worry: all we need to know is that the bigger a number is, the bigger is its logarithm, so if the number of atomic arrangements is at a maximum, so is the entropy. When the entropy of a gas (or indeed any other substance) with a fixed amount of energy has reached its maximum value, no further change will occur as long as the system remains isolated from its surroundings; the system is then in what is known as its *equilibrium state*.

Consider what will happen if more energy is added to the gas so that, on average, the atoms move faster. The range of possible values of the speed will also have increased, which means there are more ways in which the atomic velocities can be arranged, so the entropy has also increased. By the same token, if energy is removed the entropy decreases. How can this be if the principle is that the entropy should always be as large as possible? The answer is that because total energy remains the same, the energy removed from the gas must have gone somewhere else. The resulting increase in the entropy of the environment outweighs the decrease in that of the gas.

Temperature

Everyone is familiar with temperature in everyday life: we know that fire is hot and ice is cold and that if our body temperature rises, it is a pretty sure sign that we are unwell. Most people also know that temperature is associated with the form of energy called heat: the higher the temperature, the more energy the body contains and if a hot object is placed in contact with a cold one, energy will flow from the former to the latter until both are at the same temperature. To take an everyday example, when we add cold milk to hot coffee, the whole mixture soon ends up at one intermediate temperature.

To get a better physical understanding of the meaning of temperature, let's return to the example of the gas discussed above. First, all the energy is in the form of the kinetic energy possessed by the moving atoms; if more energy is added to the gas, the atoms move faster and the gas gets 'hotter'. That is, the average energy per gas atom increases. Suppose we have two identical quantities of gas where one is hotter than the other. Consider what happens if the two are put into contact, allowing energy to be exchanged between them. From our earlier discussion, we would expect that energy will flow from the hotter gas to the colder until they reach an equilibrium state where both have the same average energy per atom. It follows that, in this example, this average energy has properties very similar to that of temperature: it has a larger value when the gas is hot, smaller when it is cold and the same value when the two have reached a steady state following a free exchange of energy. Temperature is another example of an emergent property: it is pretty meaningless to say that a single atom is at some particular temperature but when a large number of atoms come together, the concept acquires meaning and usefulness.

Average energy per atom is quite a good measure of temperature in the case of a gas but it does not straightforwardly generalize to other cases, such as solids and liquids. To progress further, consider two bodies that are initially at different temperatures and are placed in contact, but completely isolated from the rest of their environment. We would expect energy to flow from the hotter to the colder, until they come to the same temperature, when no more change occurs. Remembering the earlier discussion where we saw that an isolated system containing a large number of atoms will always move towards a state of greatest disorder – that is, one where the entropy is highest – we should expect the combined system to move towards an equilibrium state where the total entropy has reached its highest possible value. Is there a property like temperature that was different for the two bodies before this final state is reached but the same for both of them afterwards? As energy flows from the hot body to the cold body, the entropies of the two bodies also change – that of the hot body decreases, while that of the cold body increases. For any given amount of energy transfer, the entropy increase must be greater than the entropy decrease because the total entropy of the two bodies combined must increase. This process continues until the total entropy reaches its maximum value at which point the entropy no longer changes significantly if a small amount of energy is transferred from one body to the other in either direction.

This idea leads directly to a definition of temperature. A small increase or decrease in the energy contained in a body causes a corresponding change in its entropy. If this energy change is divided by the entropy change and the result represented by the symbol *T*,[2] then when a given amount of energy flows from the hot to the cold body, the value of *T* for the former must be larger than that for the latter. This is because, as discussed above, the entropy increase of the former is smaller than that of the latter and the energy changes are equal in size, though of opposite sign. However, when the total entropy of the two bodies has reached its maximum value, this will not change when a small amount of energy is transferred in either direction, which means that the two values of *T* are now the same. It follows that *T* has the same properties as those we have attributed to temperature: larger for hot bodies than for cold and identical when the two have reached equilibrium. In the particular case of gases, it can be shown that *T* is just the average energy per atom, as discussed above.

In everyday life, temperature is measured by a thermometer, which may be a glass tube partly filled with a liquid such as mercury, or an electronic device. Practical thermometers are always based on physical properties that are known to depend on temperature; in the mercury thermometer, this is the volume of mercury in the tube. Another thermometer that acts on a similar principle consists of a given mass of gas held at constant pressure. The earlier discussion can be extended mathematically to calculate numerical values for *T* in this case as well as corresponding values for the volume of the gas. We find that the volume of the gas changes in direct proportion to the value of *T*. Thus if we place an object in contact with a vessel containing a known mass of gas and wait for the two to come to equilibrium, a measure of the

[2] Readers familiar with calculus will realize that *T* is the rate of change of the energy relative to entropy or the differential of energy with respect to entropy.

volume of the gas will lead to a value for the temperature of the object. This process can be used to calibrate another device, such as a mercury thermometer, which in turn can be used as a general purpose instrument.

The actual units used to measure temperature (for example degrees Celsius) are chosen to be consistent with the standard units for entropy and energy. The 'zero' of temperature is also often chosen according to convention (for example, the freezing point of water in the Celsius scale). However, looking again at the way the gas thermometer operates, we see that the zero of temperature would be better defined as the point where the volume of the gas vanishes. In fact, all gases turn to liquid before they reach this temperature but it can be shown that if this did not happen, the gas volume and the value of T would become zero at the 'absolute zero' of temperature, which is equivalent to –273.15°C in conventional units. In other words, the zero of the Celsius scale of temperature has been chosen to be equivalent to $T = 273.15$K, where 'K' stands for the Kelvin unit of temperature, named after Lord Kelvin, whose work in the University of Glasgow in the mid-nineteenth century played a key role in developing our understanding of temperature. One Kelvin (or 1K) has the same magnitude as one degree Celsius but the zero of this scale is the absolute zero of temperature.

Solids and liquids

I have discussed gases in some detail because they are simpler than the other common states of bulk matter, namely solids and liquids. In both the latter cases, the atoms are much closer together than they are in a gas, so they constitute examples of what is sometimes called *condensed matter*. Liquids are actually more difficult to understand in detail than either solids or gases, so I shall start with solids.

First, consider those solids composed from individual spherical atoms, such as are formed when a rare gas is cooled to the point where it liquefies and then further until it freezes. In the previous section it was assumed that there were no interactions between the atoms, apart from the fact that they behaved as hard balls when they collided. This is actually an approximation that is usually perfectly OK in the case of gases, but a solid relies on the fact that the atoms actually attract one another. This is true even for rare gas atoms, though the attraction in this case is very weak – about a hundred times weaker than a typical chemical bond. When the atoms come close together, they essentially 'bump into' each other, causing a repulsive force that counters the attraction. At a particular interatomic separation the repulsive and attractive forces balance and the interaction energy has its minimum value. Once this state has been reached, with the atoms at rest, they will remain joined together unless and until additional energy is supplied to drive them apart.

Suppose we bring together a large number of spherical atoms with little or no kinetic energy, so that they form a solid. The configuration with the lowest energy corresponds to an atom being close to as many other atoms as possible. If the atoms are confined to a plane, this happens when each atom is surrounded by six others, as shown in Figure 3.2(a). A three-dimensional solid often consists of a set of such planes stacked on top of each other. To get as close to the others as possible, the planes position themselves so that the atoms occupy points directly above the 'holes' in the first layer, as is shown in Figure 3.2(b). When this process is complete, each atom is surrounded by twelve others. This arrangement of atoms is an example of what is known as 'close packing' of spheres, where a set of ball-shaped objects are placed as close together as possible. An everyday application of the same principle is when a greengrocer packs as many apples as she can into a box. Close-packed structures are a common feature of many solids – particularly in the case of the elements,

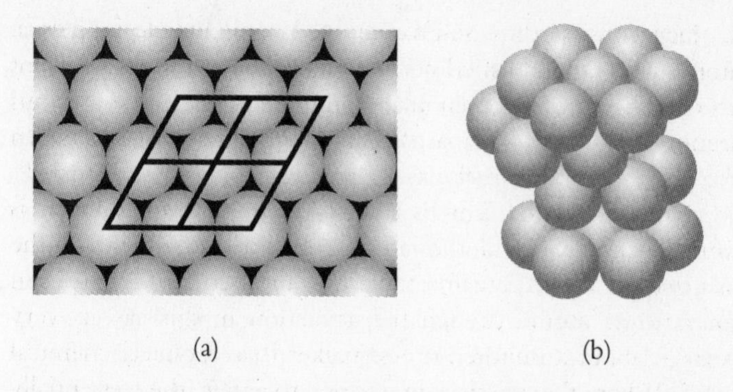

Figure 3.2 Spherical atoms confined to a plane form the close-packed arrangement shown in (a), which also shows the outlines of four unit cells. The three-dimensional close-packed structure formed by placing such planes on top of each other is illustrated in (b). The centres of the spheres in one plane lie directly above the holes in those that lie above and below it. (David Berger)

where the basic units are spherical atoms. Other substances (such as ice, which we will discuss shortly) where the basic units are molecules, adopt a wide variety of different and generally more complex arrangements.

A feature common to many different solids is also illustrated in Figure 3.2(a). This is that the solid is made up of a very large number of repeats of a basic unit: in the case of a two-dimensional sheet of close-packed spheres this building block has the shape of a rhombus, whose internal angles are 60° and 120°, with an atom on each corner. If we move this unit in a direction parallel to one of its edges by a distance equal to the length of the edge and keep on doing this in the two directions defined by these edges, we will generate the whole sheet. A similar process in three dimensions generates the atomic structure of the whole solid. The pattern of atoms created in this way is termed the *crystal structure* of the solid and the basic repeat unit is known as its *unit cell*. The word *crystal* is used because in many cases, such a structure produces a solid specimen with properties normally associated with crystals. That is, they have flat faces and sharp edges, whose orientations are determined by the atomic crystal structure. In fact, many of the solids that we encounter in everyday life are composed of crystals, though in many cases (particularly metals) most specimens are made up of a large number of small, differently oriented crystals fused together. A more detailed study of the crystal structure and other properties is able to account for the wide variety of properties that solids possess, such as the brilliance of a crystal of diamond or the ability of solid copper to conduct electricity in a very efficient manner.

An important feature of this regular structure is that the atoms are all stationary and regularly arranged; there is therefore only one possible arrangement of the atoms, so the entropy of the solid is essentially zero. Consider now what might happen if some energy is injected into the solid. There are two places where this energy might go: it might act to reduce the binding energy holding the atoms together, which will result in them moving either further apart or closer together (remember that the stationary position corresponds to a balance between the attractive and repulsive forces) or it might be given to the atoms as kinetic energy of motion. In fact, both of these processes occur together. To understand this, first consider a molecule made up of two atoms. If we add a small amount of energy to this system, it will cause the bond between them to stretch and gain energy but once all the energy has been absorbed in this way, the attractive force will cause them to move back toward the stable position. Because they are moving, the atoms will not stop there but continue until all the energy is used to compress the bond, when the atoms will again come to rest for an instant before moving apart once more. A close analogy is the motion of two balls connected by a spring. The upshot is that the atoms will be set in vibration, with the extra energy oscillating between the kinetic energy of motion and the potential energy of the bond.

The same principle applies in a solid: adding energy causes the atoms to vibrate so that the distances between them stretch and contract rapidly. In fact, this motion is best described in terms of waves of vibration with different wavelengths and amplitudes. At any one time, the atoms will be displaced from their equilibrium positions by different amounts and will be moving with different speeds. This means that the positions and velocities of the atoms are at least partially randomized and the entropy of the solid has increased. Similarly to gases, the increases in energy and entropy are associated with a rise in the temperature. An additional feature of this motion is that it is generally easier to pull atoms apart than it is to push them together, so when the atoms are vibrating, the average distance between them is a little larger than their equilibrium separation. This explains the phenomenon of *thermal expansion*: if the temperature of a solid is raised, the size of the vibrations and therefore the average distance between the atoms increases, so making the whole solid a little bit larger.

Returning briefly to the example of the molecule made up of two atoms, let us consider what happens as more and more energy is added. The atoms must vibrate with greater and greater amplitudes until the point is reached where the kinetic energy exceeds the total energy associated with the bond between them. They are then no longer attached to each other and are free to fly off in different directions. Something similar happens in the case of a solid: the vibrations become so large that at least some of the atoms are no longer bound to the others. To understand this process more fully, we apply the same principle that we used when discussing the properties of gases: when we supply energy to a solid, we expect it to be shared between the different atoms statistically so as to maximize the entropy, in which case the atoms have a range of kinetic energies. Some fraction of the atoms will have kinetic energies large enough that they are no longer bound to the rest of the solid and will be free to fly off, forming a gas. In fact, very few substances make a direct transition from solid to gas as the temperature is raised but first go through an intermediate liquid state; for simplicity, I am ignoring this for the present. (An example of a substance which does not form a liquid is carbon dioxide; the solid, known as 'dry ice', *sublimes* directly from solid to gas.)

In any given time period, there is a probability that an atom, following collisions with the other atoms in the gas and/or the walls of the container, will return to the solid surface with a lower energy than it had when it left and be reabsorbed in the solid. This results in a balanced situation, where atoms are leaving and rejoining the solid at the same rate. This is the equilibrium state corresponding to the temperature of the system. However, as the temperature rises further, the number leaving the surface increases and the chances of them losing enough energy in collisions so that they can be taken up again by the solid reduces. At some point, the number leaving the solid exceeds the number returning, the solid state is no longer stable and the material vaporizes completely. It follows that there is a particular temperature, known as the *sublimation temperature*, below which the solid and gas coexist and above which only the gas is stable. If more energy is added, this is used to convert more solid into gas and the temperature stops rising until this process is complete. The energy required to convert a given quantity of solid into gas in this way is known as the *latent heat*.

Up to now, we have barely mentioned 'pressure' and have implicitly assumed that there is a fixed body of gas in contact with the solid. However, we know that if we increase pressure on a gas, its volume is reduced so that its density is increased. If this happens to the gas in contact with the solid, the chance of the gas molecules colliding with the solid surface is increased, while the number leaving the solid is largely unaltered. More energy is needed if we are to restore the balance between leaving and arriving, so the sublimation temperature is raised.

Liquids

Most solids turn into liquids before becoming gases but what are liquids, how do they form and why do they exist? All common materials are solid if the temperature is low enough and gaseous at sufficiently high temperature. Most exist as a liquid over an intermediate range of temperatures: ice, water and steam constitute an obvious example. To further understand the process of melting, consider again the example of a solid formed from close-packed atoms (Figure 3.2) at low temperature. Because the atoms are held together by the forces between them, if we try to slide one plane of atoms across another, we would first have to pull the two sets of atoms apart. This would involve supplying energy, so raising the temperature and causing the atoms to vibrate. During this vibration, neighbouring atoms would move alternately towards and away from each other. If the kinetic energy becomes large enough, the atoms will move far enough apart for the planes of atoms to slide past each other, even though there is insufficient energy to separate them completely into a gas. This state of matter, where atoms are bound together but can slide across each other, is what we call a liquid. Normally, the average distance between the atoms is greater in the liquid than the solid, so that the former is less dense than the latter.

Evaporation – the transition from liquid to gas – is quite similar to sublimation. For an atom in a liquid to become free, it must be supplied with enough energy to overcome its attraction to the rest of the liquid; if the gas is held in contact with the liquid, the atoms will have a probability of returning to the liquid and being recaptured. If, however, the temperature is high enough, the number leaving will exceed the number arriving and the liquid will *boil*. The value of the boiling point will depend on pressure, for the same reasons we discussed in the case of sublimation. This is why it is said that good tea cannot be brewed at the top of a high mountain, because the boiling point of water is lower as a result of the reduced pressure.

Turning now to melting, many features of this transition are similar to those applying when a liquid boils to form a gas. For a given substance, the transition occurs at a particular fixed temperature and energy must be supplied before it can be completed. There is also some pressure dependence but this is usually much less than in a typical liquid–gas or solid–gas transition. In general terms, to melt a solid, sufficient energy must be supplied to pull the planes of atoms far enough apart to allow them to slide past each other When this has happened, the atoms have considerably more freedom of motion, which allows them to adapt to a comparatively random configuration, consistent with the constraint of remaining a liquid. From time to time, the sliding planes will lose energy and become attached once more but at high enough temperature, this process will be less likely than the one that causes the planes to separate and slide. In this case, the liquid will be the only stable form; the lowest temperature at which this occurs is known as the *melting point*. Just as in the case of the solid-to-vapour transition discussed above, the liquid and solid states will coexist at the melting point: adding more energy converts some of the solid into liquid, without raising the temperature until this process is complete, so that latent heat is also a feature of the melting transition.

The existence of the three states of matter is therefore a direct consequence of the properties of atoms and molecules, which are themselves the results of the interactions between the electrons and nuclei contained in the atoms along with the fundamental laws of quantum physics. Real and important properties of materials, such as solidity and liquidity, emerge naturally as a consequence of the coming together of their fundamental constituents. Reductionism, as defined and discussed in earlier chapters, therefore still applies. Because such ideas have been embedded in our culture for so long, this idea seems quite obvious and unsurprising but it is actually quite amazing that the physical laws that apply on the microscopic, subatomic scale are all that are needed to explain the variety of properties displayed when large numbers of atoms are brought together.

Water and ice

The traumatic events that affected the Shackleton expedition were due to the contrasting properties of liquid water and solid ice. Water is one of the commonest substances on this Earth and is an essential part of all known living systems. The word *water* can either be used specifically to describe the substance in its liquid form or as a generic term for the substance whatever its state – solid (ice), liquid, or gas (steam). This section aims to explain how the physical attributes of these three forms emerge naturally from the basic properties discussed in Chapter 2. You will recall that the water molecule consists of three atoms: one of oxygen and two of hydrogen (hence the formula H_2O). Each hydrogen forms a bond with the oxygen and the two OH bonds form an angle of about 105° – see Figure 2.3. Each bond is formed by sharing electrons between the oxygen atom and one of the hydrogens. A further consequence is that the electron that surrounded the isolated hydrogen atom in the form of a spherical cloud is pulled towards the oxygen when the bond is formed. As was discussed in Chapter 2, this means that the positive charges on the hydrogen nuclei are not fully cancelled by the negative charge of the surrounding electrons, while the cloud surrounding the oxygen atom contains somewhat more negative charge than is possessed by the positively charged oxygen nucleus.

Consider what happens when two water molecules are brought together to maximize the electrical attraction between the positive and negative charges. They will try to arrange themselves so that a hydrogen atom in one molecule is close to the oxygen atom in the other and away from the hydrogen atoms associated with it. Such a connection between the two molecules is known as a *hydrogen bond*. This has appreciable strength, although it is more than ten times weaker than the chemical bond between an oxygen and a hydrogen atom in the same molecule. Figure 3.3 illustrates the arrangement of molecules in solid ice. Each oxygen

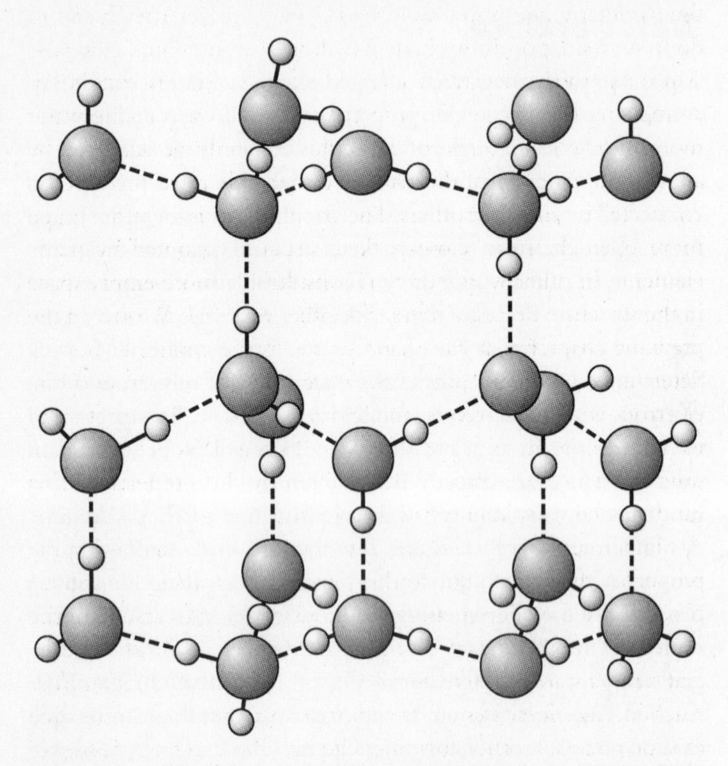

Figure 3.3 A perspective sketch of the arrangement of water molecules in the crystal structure of ice. The larger grey spheres and the smaller white spheres represent oxygen and hydrogen atoms respectively. The solid lines show the bonds connecting the oxygen and hydrogen atoms within a water molecule, while the broken lines indicate the connections between positively charged hydrogen atoms and negatively charged oxygens in neighbouring molecules. The structure is relatively open when compared with that of close-packed spheres – see Figure 3.2. (David Berger)

atom is connected to only four other oxygen atoms, with a hydrogen atom lying near the line joining two oxygens; this contrasts with a close-packed structure, in which each atom has twelve nearest neighbours. This arrangement is a direct result of the fact that when water molecules come together, they prefer to do so with the positively charged hydrogen atoms coming as close as possible to the negatively charged oxygens. Because the bonds connecting the oxygen atom to the two hydrogens in the water molecule are at an angle of 105°, this can only be achieved by adopting a structure of the type described, where each oxygen is connected to only four others. The arrangement is therefore much more open than the 'close-packed' structure adopted by many elements. In other words, there is considerably more empty space in the structure of ice than in many other materials. We saw in the previous chapter that the shape of the water molecule is itself determined by the quantum rules governing the properties of the electrons within the oxygen and hydrogen atoms. The structure of ice is therefore an example of how the essential properties of this solid substance are directly determined by the properties of its fundamental constituents and the physical laws governing them.

The melting of solid ice into liquid water displays some properties that are unique to this substance. Applying the principles discussed earlier, if energy is added to ice, this results in the molecules moving apart, along with an increase in the entropy and temperature of the material. Once a particular temperature is reached, the molecules are far enough apart for them to be able to slide past each other, forming a liquid whose entropy is higher than that of the solid, which then melts. The separation of the molecules means that the forces holding them in position, with their hydrogens pointing towards neighbouring oxygen atoms are considerably weakened. This allows the molecules some freedom to rotate, leading to greater disorder and an increase in the entropy. However, this rotation allows the rotated molecules to approach each other more closely than is the case in the open structure described above. As a result, when ice melts, the H_2O molecules in liquid water are on average closer together than they are in ice, which means that water is denser than ice and so ice floats on water. Because water is by far the most commonly encountered example of melting, it could be thought that this behaviour is typical of any melting process, but in fact, water is one of very few substances where the solid form is less dense than that of the liquid. More typical melting behaviour can be observed in substances such as candle wax: when a candle burns, liquid wax typically forms a pool at the base of the wick; when the candle is extinguished, the liquid freezes from below to form part of the solid wax.

Further consideration of this process reveals another important property, which is probably unique to water. Although most of the orientation-preserving connections between the molecules are broken on melting, some remain, so the liquid just above the melting point contains regions where some of the open structure is preserved. These continue to break down as the temperature is raised, allowing the density of the water to increase further until, at about four degrees above the freezing point, the normal processes of thermal expansion dominate and the density reduces as the temperature is raised further. The fact that water expands when cooled below 4°C is described as the *anomalous expansion* of water.

These properties of water play a vital role in ensuring that the physical environment of our planet is as we know it. Otherwise, a drop of the surface temperature towards 0°C would cause cold water and then ice to fall to the bottom of a sea or lake. It would be very difficult to reverse this process, as a rise in temperature at the surface would result only in a warming of the surface layers of the water. Over a series of seasons the ice layer at the bottom of the sea would grow in the winter but would be little affected by the warmer summer weather. Eventually, apart from a comparatively thin layer at the surface, the whole sea would be frozen.

As a result life on Earth as we know it could not be supported, and indeed, might well not have evolved. Sometimes this fact is described as 'miraculous' because without it we would not be here. But this is a very Earth-centred view. Planets elsewhere in the universe could be effectively identical to Earth but have a climate that is just a little bit warmer, so that the temperature never falls far enough for water to freeze.

In any case, no miracle, in the sense of a departure from the laws of physics, is involved, because the properties of water and ice are determined by these laws, which have the consequences that:

- An oxygen atom is composed of a nucleus carrying eight positive charges surrounded by eight electrons. Two electrons are in spherical states and four are in higher energy states, whose dumbbell-shaped wavefunctions are mutually perpendicular.
- When two hydrogen atoms are brought close to an oxygen atom, the electron associated with each combines with a single oxygen electron in one of the orbitals to form a chemical bond.
- The resulting molecule of H_2O has an H-O-H angle of about 105°.
- The electrons in the H_2O molecule are more likely to be found on the side of the oxygen atom opposite to the hydrogens, which therefore carry a net negative charge, while the hydrogen atoms carry a net positive charge.
- When many H_2O molecules are brought together at low temperature, the positive and negative charges on neighbouring molecules attract each other and a rigid, but quite open, structure is formed.
- When energy is supplied to raise the temperature to 0°C, the molecules become detached to the extent that they can slide past each other, forming a liquid.

- Once this happens, some of the molecules are no longer held in a rigid open structure, so they can move closer together, resulting in water being denser than ice.
- As the temperature is raised a further few degrees above the melting temperature, the fraction of liquid with this open structure reduces.

Thus, the wetness of water, the hardness of ice, the unusual features of the process of melting and anomalous expansion are direct consequences of the laws that determine the behaviour of the fundamental particles that make up the hydrogen and oxygen atoms. Indeed, if the properties of water were different – if, for example, ice were denser than water, this would actually be inconsistent with the basic laws governing the fundamental component particles. Because this falsification does not occur, the principles of reductionism continue to apply.

How does all this apply to the scenario that we began this chapter with, where Shackleton's *Endeavour* was crushed by ice in the Weddell sea? First, some of the H_2O molecules are rigidly bound to others in a solid open structure, while others have become partly detached from their neighbours and can move past each other as a liquid. The assemblages of bound molecules float on the liquid, because ice is less dense than water. Second, the temperature of the air above the sea is lower than the melting point of water, which results in more of the water freezing to become ice, although not enough energy has yet been extracted for this process to have been completed. Third, under pressure from the wind and tide, the molecules in the solid are moved through the water as blocks of ice; these blocks collide with each other and the sides of the ship, exerting pressure as they do so. The surfaces at the collision points rub together, setting the molecules in the vicinity into vibration; these vibrations are transmitted into the air and then into the ears of human bystanders who perceive them as a variety of sounds.

Thus, we can say that everything described in our opening passage is a direct consequence of the laws of quantum physics.

Despite the above, does a completely reductionist description tell us everything important about the events described? Many of us would say 'no' because it contains little or none of the drama captured in the original account, with its metaphors and poetic allusions. The 'enormous jigsaw puzzle', 'the unhurried deliberateness of the motion' leading to the impression of 'implacable power' are powerful images that tell us much more about what is happening at the level of the ice floes, the ship and the human observers than the molecular description ever could. But it is consistent with reductionism because all these happenings emerge from and supervene on the molecular-level description. Each H_2O molecule and, indeed, each electron and nucleus within it, obeys the fundamental laws of quantum physics, and these inevitably produce the large-scale phenomena that prompt such graphic description.

4

The chemistry of life

The contrast between living and non-living things is obvious to the most casual observer of nature. Indeed, the world of plants and animals is so different from the rest of the universe that it is easy to believe that these creatures and their development must be governed by completely different principles and laws. Until it was challenged by the scientific revolution that began in the seventeenth century, the idea – known as *vitalism* – of a separate set of fundamental laws for living organisms underlay most attempts to account for their behaviour.

Even today, vitalist concepts survive. We see them embedded in terms such as 'life force' and 'vital spark', which are commonly applied to living beings. Much of so-called alternative medicine is founded on the idea that the body is subject to 'energy fields', which have little or nothing to do with the concepts of energy and field in physics. A prime example of this type of thinking is the statement by Samuel Hahnemann, the founder of homeopathy, in which he described disease as due to 'solely spirit-like (dynamic) derangements of the spirit-like power (the vital principle) that animates the human body'.

In contrast to vitalism, the reductionist approach asserts that the properties and behaviour of living things should emerge from, or supervene on, those of their constituent atoms and molecules. This should be true even though the latter are governed by the fundamental laws and principles already discussed in previous chapters. Living matter is hugely more complex than anything discussed up to now, however, and a detailed explanation of all the processes involved is well beyond the reach of this book. I will therefore concentrate on one or two key areas that exemplify the reductionist principle in the context of living systems.

The previous chapter explained how the properties of a substance such as a simple gas, liquid or solid result from it being in an equilibrium state. This followed from combining the principles of minimizing the energy of interaction between the constituent atoms and maximizing their total entropy, which is a measure of the disorder of the whole system. The form (solid, liquid or gas) the substance adopts depends on its environment, in particular the temperature of its surroundings.

Living things, however, are in a state that is anything but equilibrium. In biological systems, the atoms neither occupy a regular position in a crystal, nor are they arranged randomly, as in a gas. Living matter consists of a large set of complex, highly ordered systems that interact with each other in many different ways and which are often in a state of flux. As one example, consider your blood: it is a fluid pumped through your veins by the muscular action of your heart, which is fuelled by the food you have eaten and the air that you have breathed.

It follows that an important question is how the complex structures that form living matter emerge from the reactions between comparatively simple chemicals. Very little flexibility is allowed in the way the atoms making up a living organism can be arranged without destroying its essential function. In this sense the organism is highly ordered, so what has happened to the law of increasing entropy? Doesn't reductionism require that this principle should apply to living matter just as it does to a glass of water at room temperature? To answer this, we have to understand that living matter is in a state where energy moves continually through the system in such a way that it is *kept away* from equilibrium. When this stops, the organism dies and eventually decays into a comparatively high-entropy state.

Thus, living things have to overcome the disorder inherent in other entropy-bound matter. To help understand this, consider a relatively simple example from another area of science where order spontaneously emerges from a previously disordered system. A quantity of fluid held in a vessel at a single temperature appears smooth and devoid of structure (certainly at any scale larger than that of its component atoms). Now suppose this fluid is heated from below. If this results in only a small temperature difference between the top and the bottom of the fluid, heat will be conducted through it and its appearance will remain largely unchanged. However, if the heating rate and associated temperature difference both increase, a process known as convection sets in. The fluid near the heat source expands and rises up towards the top of the fluid, where it cools and falls back down. Different parts of the fluid do not rise and fall at random but, instead, a set of *convection cells*, composed of rotating sections of the fluid, is set up. Some of these cells rotate clockwise and others rotate anticlockwise. If the experiment is performed carefully, the cells are of similar size and regularly spaced, as is illustrated in Figure 4.1. Scientists have used computers to simulate this process and found that the rotating cells emerge spontaneously, even though the calculations assume that the component atoms

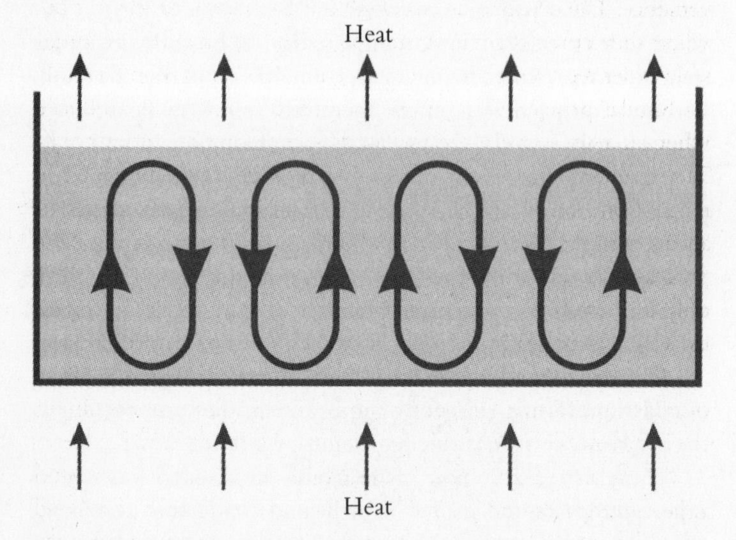

Figure 4.1 An illustration of convection cells that form in a fluid when it is heated from below. (Alastair Rae)

obey the standard laws of motion. The formation of convection cells is therefore an example of how order can appear spontaneously in a previously disordered system. It is, however, consistent with the principle of maximum entropy, because the entropy generated by the heat flowing out into the cold environment at the top is greater than the entropy flowing in from the heat source at the foot.

Another example of convection involves a planet's atmosphere responding to the heat of a sun. The Earth's atmosphere allows most of the sun's heat to pass through it in the form of radiation, which heats the surface of the Earth. This, in turn, heats the atmosphere from below, setting up convection currents, which results in the large-scale movements of the air that are known as weather patterns. These air movements are similar in principle to simple convection cells but are considerably more complex because other factors – in particular the rotation of the Earth – play an important role.

The appearance of convection cells is another example of supervenience. The separate atoms obey the basic laws of physics but when they come together in the presence of heat flow, the large-scale order represented by the convection cells arises spontaneously.

Similar principles appear in living matter. Energy is produced when an animal or plant 'burns' food by combining it with oxygen. The resulting heat powers a host of processes in the body. These range from small 'protein machines' built from a few thousand atoms to large-scale mechanisms, such as the heart or lungs. All these processes reduce the entropy of parts of the system; this is possible only because living bodies emit heat, which raises the entropy of their environment, ensuring an overall net entropy increase.

The living world regularly confronts us with dramatic examples of order appearing apparently spontaneously. Take, for example, the development of a chicken embryo inside a fertilized egg. The bulk of the material in the egg (the yolk and the white) appears featureless to the casual observer but the embryo grows and develops into a chicken, which is a highly structured object. An inevitable by-product of the chemical reactions associated with this process is the emission of heat into the environment and a consequent increase in its entropy.

The living cell

The basic unit of all living matter is the *cell.*[3] Cells were first observed by the English scientist Robert Hooke, around the middle of the seventeenth century. Hooke was one of the outstanding scientists of that period; he was accomplished in a wide variety of fields, including physics and astronomy as well as biology. His reputation would probably be much greater if he had not been overshadowed by Isaac Newton, who lived at the same time. Indeed, they were fierce rivals, to the point where Hooke claimed he had invented the theory of gravity before Newton. Hooke discovered cells when he examined thin sections of cork through a microscope he had developed himself. He gave them their name because their appearance reminded him of the shape of the cells used by monks in monasteries.

It is now known that cells come in a variety of shapes and sizes, suited to the function they perform. For example, skin cells join together to form protective coatings and nerve cells carry electrical impulses to muscle cells, which then collectively stretch and contract. A huge variety of cell types are needed to perform the specialist tasks that are essential to the existence and operation of living bodies.

A typical cell is about one millionth of a metre across – small in everyday terms but about ten thousand times larger than an atom. Most cells consist of two main regions: the *cytoplasm*, which constitutes the main volume of the cell and the *nucleus*, which is a comparatively small region housed within it. (This should not of course be confused with the atomic nucleus discussed

[3] Biological cells are, of course, quite different from the convection cells discussed above or the unit cells discussed in Chapter 3.

in Chapter 2.) The whole volume is enclosed by the *cell membrane*, which has openings that allow water and other material, such as sodium ions, to pass in and out.

The cytoplasm is primarily made up of protein molecules. A typical protein molecule contains around a hundred thousand atoms, which means that it is much larger than any of the molecules discussed so far. A typical cell contains thousands of different types of proteins, each of which plays a role in the cell's operation. There are proteins that can bind to each other to form filaments and tubes and others that join together to form the sheets that make up the cell's outer membrane. Particular proteins, known as enzymes, interact with other molecules and break them down into smaller components. For instance, the action of an enzyme on a sugar molecule causes the latter to split into its component atoms, which then react with oxygen to produce the power needed for the multitude of reactions involved in the cell's operation.

Centre stage in the cell's nucleus is the molecule known as DNA, which contains the genetic information used to manufacture the various proteins needed in the cytoplasm. The DNA molecule is also able to reproduce itself by dividing into two identical parts. All the cells contained in an organism contain essentially identical copies of its DNA.

How the relevant pieces of genetic information come together to construct particular proteins in particular cells is a very complex process. Compared with this, the structure and reproduction of DNA molecules are relatively simple topics and these are the examples I shall concentrate on.

The structure of DNA

DNA is an abbreviation for *deoxyribonucleic acid*. The average molecule of DNA contains hundreds of millions of atoms but its structure is essentially simple. It is made up of two long, quite complex, chains of atoms (known as 'backbones'), to which, at regular intervals, are attached additional molecular structures each made up of between ten and twenty atoms. These *nucleotides* or *bases* come in four types: adenine (A), cytosine (C), guanine (G), and thymine (T). The letters in brackets are handy abbreviations that are used far more frequently than the full names. The DNA bases can form connections with each other, provided they form pairs composed of either adenine–thymine (A–T) or cytosine–guanine (C–G). The links between these base pairs hold the two backbones together. Although the bases can appear in any order along one or other of the backbones, the restrictions on the possible pairings ensure that the sequence on one backbone has to complement that on the other, as illustrated in Figure 4.2.

The connections between the bases in a pair are appreciably weaker than the bonds between a base and a backbone and this is essential to understanding how a DNA molecule reproduces. Suppose that some DNA is immersed in a medium that contains the components needed to form a new DNA backbone, including a supply of the molecules corresponding to the bases. Because the inter-base connections are relatively weak, there is a significant chance of them separating over a small section of the molecule; this may happen at the end of a DNA chain but is more likely to occur somewhere in the middle. Once it occurs, it opens up the possibility of the unbound bases pairing up with new bases, which, in turn, can bond with the molecules needed to form new backbones, as shown in Figure 4.3. The bases used in this process are identified and supplied by the action of a protein, polymerase. As the splitting proceeds, pressure on neighbouring sections of the original molecule is increased, so they, in turn, can form more new sections of backbone with their base pairs. Quite quickly, the whole of the original molecule 'unzips' and two molecules of DNA are formed. Moreover, because of the restrictions on how bases can be paired, the order of the base pairs is identical in the two molecules. In other words, exact reproduction has taken place.

Once the DNA in the nucleus has reproduced itself in this way, a complex series of actions is triggered. These involve proteins

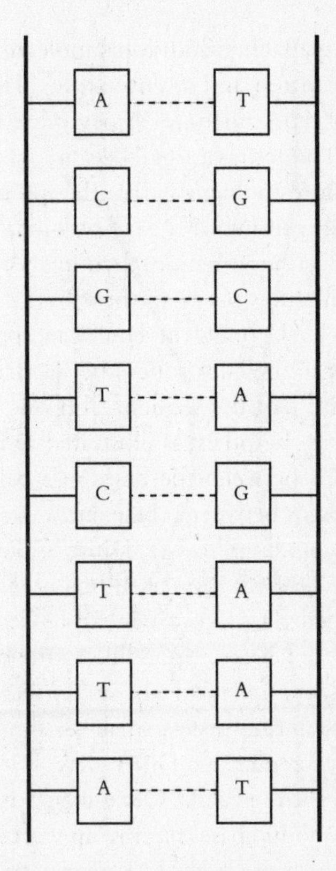

Figure 4.2 Schematic representation of a short section of a DNA molecule. The vertical lines represent the two backbones and each square box stands for one of the bases. Base pairings are indicated by the broken lines. (David Berger)

in both the cytoplasm and the cell wall. The eventual result is the division of the whole cell into two copies of itself, each with a nucleus containing one of the identical DNA molecules. This type of cell division (known as *mitosis*) plays an important role in maintaining the organism by replacing cells that have become damaged or missing, such as in the healing that takes place after a cut in the skin. Important mechanisms are in place to ensure that

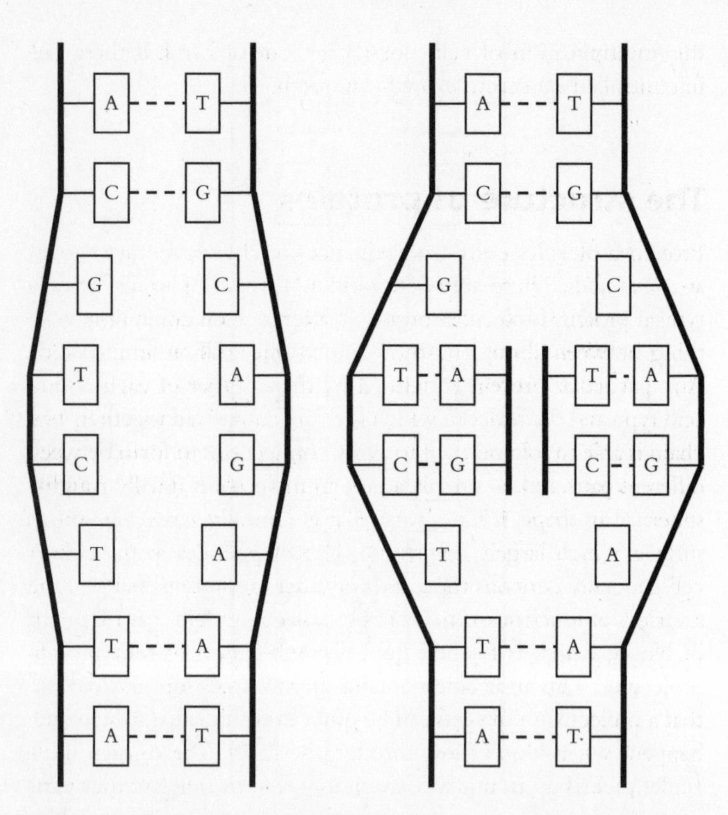

Figure 4.3 The left-hand diagram represents the section of DNA illustrated in Figure 4.2 but with the connections between the central four base pairs broken. In the right-hand diagram, two of these broken pairs have joined up with fresh sections of backbone and new bases to form two new sections of DNA that are identical with the same section of the original. This produces more splitting pressure on the chains to split apart, so this process continues until two copies of the whole of the original DNA molecule are produced. (Alastair Rae)

this multiplication of cells doesn't get out of hand; if these fail, uncontrolled cancerous growth can result.

The structure of proteins

Protein molecules consist of sequences of chemical units known as *amino acids*. There are twenty different types of amino acids; a typical protein molecule consists of a long chain containing anything between about one thousand to one million amino acids. Any particular protein is defined by the number of each amino acid type and the order in which they are connected together. The chain is able to fold on itself to allow connections to form between different parts and, as a result, a protein molecule is usually roughly spherical in shape. It has a particular chemically active area on its surface, which largely determines how it operates in the cell. A cell generally contains thousands of different proteins performing a variety of functions. One example is haemoglobin, a component of blood, which is formed from a combination of four protein molecules. This molecule contains an atom of iron arranged so that a molecule of oxygen can be quite easily attached to it, which happens when blood flows through the lungs. The oxygen molecules picked up in this way are transported to cells in other parts of the body, where they are combined with hydrocarbons and/or carbohydrates to release the energy needed for cells to operate. For example, muscles use this energy to produce movement.

Which particular operations a given protein performs is determined by the order of its amino acids; which in turn is determined by the information contained in the cell's DNA. The process whereby proteins are manufactured using the information contained in DNA proteins is complex in detail but simple in principle. The order of amino acids in a given protein molecule is determined by the order of the base units in a particular section of DNA – but remember that there are only four different types of base unit, while there are twenty different amino acids. This problem is overcome by the use of a three-letter code, in which every amino acid is associated with a series of three bases in the DNA sequence. For example, if the DNA molecule contains the sequence CTA, the amino acid known as leucine is added to the protein chain, while the sequence CGA corresponds to the amino acid cysteine. There are sixty-four possible triplet combinations of the four varieties of base pairs contained in DNA; all are used in the process of building protein molecules. Some amino acids are coded by more than one triplet, while some specific triplets tell the DNA to begin the manufacture of a protein and others to finish it.

This process requires the amino acids to be present in the cell, so that they can be used to construct the protein molecule. Some amino acids are manufactured in the body, but others have to be provided from an external source. These latter originate in the gut from the breakdown of proteins that have been eaten as food; they are then transported to the cell via the blood stream. This is why proteins are an essential part of a healthy diet.

A sequence of bases in a DNA chain between a start triplet and a stop triplet is known as a *gene*. Every gene (which often contains several thousand base pairs) contains the information needed to produce a protein with a particular sequence of amino acids, which in turn determines the properties and mode of operation of the resulting protein molecule. All the cells of a given organism contain identical DNA in their nuclei. This means that the information in the DNA sequence can be used to produce the proteins appropriate to a particular cell – be it in the organism's skin, heart muscle, eye, or whatever. This is achieved by the relevant genes switching on as a result of the action of certain proteins that are already present in the cell. These act on the DNA molecule to open it up and unwind it to the start point of the gene.

Although the DNA is contained in the cell's nucleus, protein manufacture occurs mainly in the cytoplasm. Information transfer involves another molecule, known as *messenger RNA*. Like DNA, this is composed of four types of bases connected to a backbone but, unlike DNA, it consists of only a single backbone. A molecule of messenger RNA is built by copying the coding sequence for the gene from one of the DNA strands until a stop sequence is reached. At that point, the copying ends and the completed RNA molecule moves out into the cytoplasm. There it encounters what is known as a *ribosome*, which is composed of a number of proteins that act together as a protein machine. This machine detects the sequences of triplets on the messenger RNA to select the relevant amino acids, which are then connected together to form the protein molecule.

This is only a brief précis of the complexity of protein manufacture. As well as the information contained in the DNA, the process depends on the proteins necessary to construct the machines existing in the living system. These proteins select the relevant gene and use it to identify the sequence of amino acids that constitute the protein that is being manufactured. They have to be created when the cell is first formed, which may have been the result of a division of earlier cells of the same type or may have been generated as part of an organism emerging from a fertilized seed or egg. Where the first of these appeared from is another question, which we shall touch on a little later.

So far, the molecules of both DNA and proteins have been described in rather general terms. However, the DNA backbone and bases, along with the amino acids which form the building blocks of proteins, have the form of molecules composed of atoms joined together by chemical bonds. The principles of chemical bonding were introduced in Chapter 2; the next few sections show how they can be used to develop an understanding of the structure and properties of DNA and proteins at the atomic level.

Carbon and hydrocarbons

An essential component of all living matter – at least as experienced on Earth – is the presence of the element carbon. Indeed, students of chemistry are often taught that their subject consists of two parts: 'organic chemistry', which is the study of compounds containing carbon and 'inorganic chemistry', which is the study of all the others. There must be something special about carbon.

We saw in Chapter 2 that an atom consists of a small, positively charged nucleus surrounded by a number of negatively charged electrons. The electrical attraction that keeps the electrons in the vicinity of the nucleus and their quantum states is described by wavefunctions that are determined by the Schrödinger equation. The lowest energy state has a spherical wavefunction that peaks close to the nucleus and the next highest energy level is associated with four different wavefunctions. One of these is spherical, while the other three have a dumbbell shape, with their long axes at right angles to each other (see Figure 2.2). The Pauli exclusion principle requires that no more than two electrons can be associated with any single quantum state.

A carbon atom contains six electrons. Two of these occupy the lowest energy state, leaving four to be distributed among the four wavefunctions associated with the first excited state. One might expect this to be achieved by assigning one electron to each of the four excited states or perhaps have two associated with the spherical state and the other two distributed between the dumbbells. In fact, when a carbon atom forms bonds to other atoms, an additional principle, known as *hybridization*, comes into play. Because the four excited states have nearly equal energies, the Schrödinger equation allows them to be combined to produce four states that have similarly shaped wavefunctions but with different orientations. These are arranged symmetrically, as shown in Figure 4.4(a). This arrangement can be described by imagining a tetrahedron (which is a symmetric four-sided figure) with the nucleus at its centre and the lobes of the wavefunctions pointing towards each vertex.

These properties of the carbon atom allow the formation of molecules known as *hydrocarbons*. In Chapter 2 we saw that if two atoms approach each other, their wavefunctions can combine to

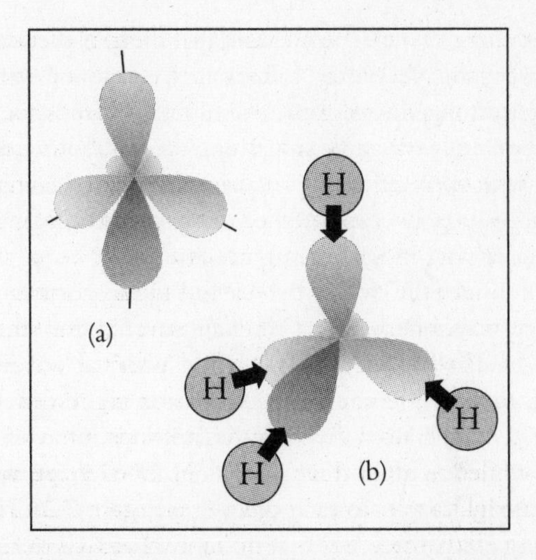

Figure 4.4 (a) The wavefunctions of the carbon atoms have the form of four symmetrically arranged lobes in a tetrahedral configuration. Each lobe has one electron associated with it. (b) When each of these lobes overlaps with the wavefunction of a hydrogen atom, a molecule of methane is formed. (David Berger)

produce a lower energy state, which can then be occupied by one electron taken from each atom; this results in chemical bonding. For example, a hydrogen molecule is formed when two hydrogen atoms come together and share their electrons in this way and a water molecule is formed when hydrogen wavefunctions overlap with two of the dumbbells associated with oxygen, to form hydrogen–oxygen bonds. These principles can also be applied to understand how carbon and hydrogen can combine to form the class of molecules known as hydrocarbons.

Consider the case of the hydrocarbon methane. This molecule consists of a carbon atom surrounded by four hydrogen atoms in a symmetric, tetrahedral configuration, as shown in Figure 4.4(b). Each hydrogen atom wavefunction overlaps with one of the lobes of the hybridized carbon wavefunctions, forming a C–H bond containing two electrons. Methane (CH_4) is therefore quite a stable molecule. Nevertheless, its energy, combined with that of two oxygen molecules, is greater than that of four molecules of water and one of carbon dioxide, which means that methane can combine with molecules of oxygen in the air, releasing energy. Natural gas is largely composed of methane and this is the basic process underlying its use by humans as a fuel.

Now consider the case of two carbon atoms coming together so that they are connected by a bond constructed from the overlap of two hybridized lobes, one associated with each atom. There are then three lobes remaining on each atom, each of which can bond to a hydrogen atom, creating the molecule known as ethane, which is represented by the symbol $H_3C–CH_3$. Alternatively, one of these hydrogen atoms could be replaced by a third carbon atom, along with three hydrogens, to form $H_3C–(CH_2)–CH_3$, which is known as propane. This process can be continued more or less indefinitely to produce the molecules of 'long-chain' hydrocarbons that have the general formula $H_3C–(CH_2)_{n-2}–CH_3$, where n is the number of carbon atoms in the chain.

The tetrahedral arrangement described above is one of two main ways in which carbon atoms can come together to form molecules. Another, less-symmetric option is one where the spherical state combines with only two of the three dumbbells to form a threefold symmetric 'star-shaped' arrangement that lies in a plane passing through the atomic nucleus. The third dumbbell is perpendicular to this plane. When two such carbon atoms are brought together, they can share two electrons within the plane and another two between the dumbbells pointing perpendicular to the plane. This produces a very strong attraction, known as a *double bond*. The other four orbitals are then available to bond to other atoms, such as hydrogen. One of the simplest molecules of this type has the representation $H_2C=CH_2$ and is known as ethene (not to be confused with eth*a*ne). It is also possible for six carbon atoms to be joined together in this way to form a hexagonal ring;

if each carbon atom is also bound to a hydrogen atom, the resulting compound is benzene. If some of the hydrogen atoms in benzene are replaced by more carbon atoms, larger and more complex hydrocarbon molecules can result. The resulting shapes of some of the molecules discussed are shown in Figure 4.5.

Whether the tetrahedral or the star-shaped bonding arrangement is adopted by a carbon atom in a particular case depends on the context – that is, the relative availability of other carbon

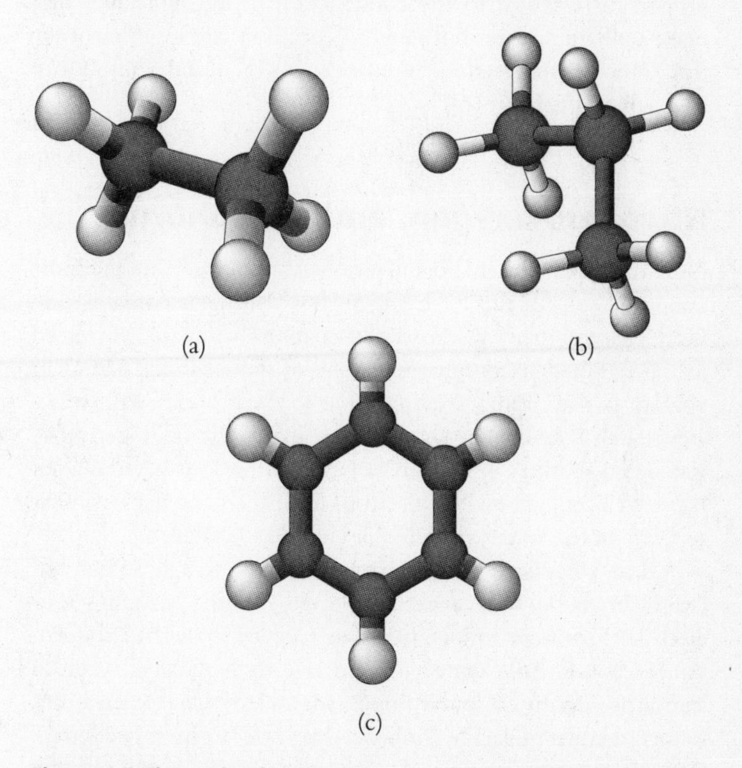

Figure 4.5 The molecular structure of the hydrocarbon molecules: (a) ethane, (b) propane, (c) benzene. In each case, the darker spheres indicate carbon and the lighter, hydrogen atoms. (David Berger)

atoms, hydrogens, and so on. In the laboratory, chemists create environments that are conducive to the formation of the type of molecule they want to study. In nature, hydrocarbons are usually found only in living matter or in matter that has once been part of a living organism: they are an important ingredient of fossil fuels, such as oil and gas.

Hydrocarbons exploit the ability of carbon atoms to bond together to form chains of more or less indefinite length. This property is essential to the construction of the molecules that make up living matter but pure hydrocarbons are insufficient for this. Other elements, notably nitrogen, oxygen, and phosphorus, have an essential part to play.

Nitrogen, oxygen, and phosphorus

Next to carbon and hydrogen, nitrogen and oxygen are the most common constituents of living matter. Nitrogen contains one more electron than carbon, so if its quantum states are hybridized in the tetrahedral configuration, one of the resulting states can contain two electrons, leaving three to form bonds with other atoms such as hydrogen. Thus in ammonia (NH_3) a nitrogen atom is bonded to three hydrogen atoms and the fourth tetrahedral hybrid is occupied by two electrons to form a lone pair – similar to those in the water molecule discussed in Chapter 2.

Oxygen has two more electrons than carbon, so it can bond to two hydrogens to form a water molecule, with the remaining four electrons forming two lone pairs (see Chapter 2). It can also form double bonds. An example of this is carbon dioxide, where a carbon atom forms a double bond with each of two oxygen atoms to form a linear molecule symbolized as O=C=O. In this molecule, four electrons are associated with each double bond and the remaining eight electrons form four lone pairs with two on each oxygen atom on the side facing away from the central carbon.

The presence of the two double bonds in carbon dioxide means that four of its electrons are in particularly low energy states. This is one reason why energy is released when a hydrocarbon is combined with oxygen to produce carbon dioxide along with water.

Another element that is relevant to building a reductionist understanding of life is phosphorus. Phosphorus has fifteen electrons. The number of electrons in the rare gas neon is ten, so phosphorus has five electrons available to form bonds to other atoms. Various bonding configurations are possible but the one that is relevant to our present discussion is known as the *phosphate group*. This consists of a phosphorus atom bound to three oxygen atoms and one hydrogen atom. One of the phosphorus–oxygen bonds is a double bond carrying four electrons; the other two oxygen atoms are each bound to a further hydrogen atom. All twenty-two bonding electrons are accounted for: four in the double bond, two in each of the other P–O bonds, two in the P–H bond and twelve in the six lone pairs associated with the three oxygen atoms.

The atomic structure of DNA and proteins

DNA is a much larger molecule than any we have discussed in earlier chapters but it consists of atoms connected by chemical bonds similar to those discussed above. As mentioned earlier, the DNA backbone consists of a large number of identical 'repeat units' joined together. Figure 4.6 illustrates the chemical bonding involved in one of these repeat units. To understand this, first consider the ring formed by four carbon atoms and one oxygen, shown in the bottom right of the diagram. The configuration of this oxygen atom is quite similar to that in the water molecule, except that it is connected to two carbon atoms instead of two hydrogens. There are also single bonds between pairs of carbon atoms and the angles are such that this five-atom ring is nearly planar. Each carbon atom forms two more single bonds to other atoms, including an additional carbon atom that is connected to a further oxygen atom and two hydrogens.

The rest of the unit consists of a phosphate group, similar to that discussed above, except that the phosphorus–hydrogen bond has been replaced by a bond to the oxygen attached to the foot of the five-atom ring in the next repeating unit. This is how the repeat units join together to form the strands that make up the DNA backbones.

A base unit is attached to the backbone at the point corresponding to the position marked B in Figure 4.6. These come in four types and their chemical structures are shown in Figure 4.7. Each base is constructed from carbon, hydrogen, oxygen, and nitrogen atoms bonded together following the principles discussed above. The four are shown in the diagram as two pairs, each composed of two molecules connected together by hydrogen bonds, as indicated by the broken lines. These are similar to the hydrogen

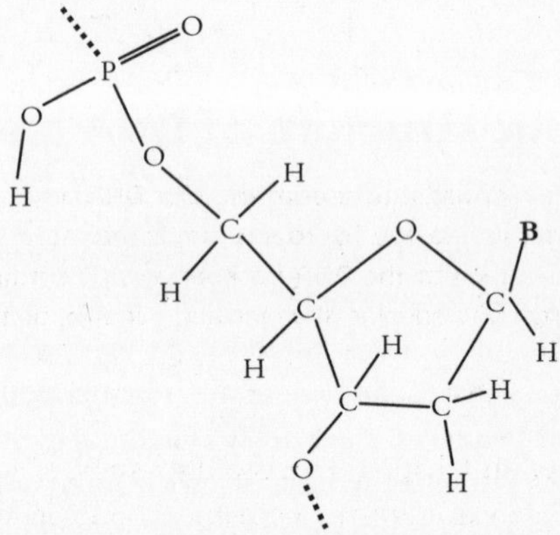

Figure 4.6 The DNA backbone consists of repeating units, with the centres of the atoms indicated by letters: hydrogen (H), carbon (C), oxygen (O), and phosphorus (P). The dotted lines coming from the phosphorus and oxygen atoms indicate bonds formed with the oxygen and phosphorus atoms respectively in the next units of the chain. The symbol B represents the point where a base unit is attached. (Alastair Rae)

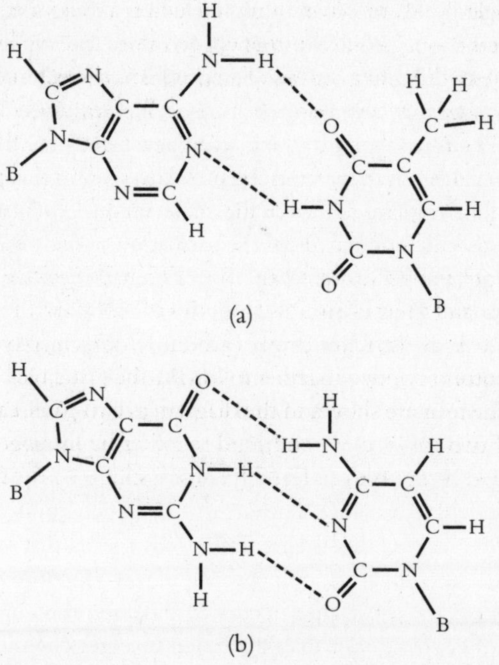

(a)

(b)

Figure 4.7 The molecular structures of the four DNA bases, showing how they pair up by forming the hydrogen bonds indicated by the dotted lines. The bases attach to the DNA backbones at the points marked B. Panel (a) shows the pair adenine–thymine and panel (b) guanine–cytosine. (David Berger)

bonds discussed in Chapter 3, in the discussion of water and ice. When a hydrogen atom forms a chemical bond, its wavefunction is changed so that the electron is more likely to be found within the bonding region than outside it; conversely, the electron is less likely to be found on the opposite side of the nucleus from the bond, so, when observed from this direction, the hydrogen atom appears to carry a small positive charge. The opposite situation arises when a lone pair is formed on, say, an oxygen or nitrogen atom: electrons are now more likely to be found in the region of the lone pair and a local excess of negative charge results. Thus, if a hydrogen atom associated with one molecule points towards a lone pair belonging to another one, an attraction results. Figure 4.7 shows that two such hydrogen bonds are formed in the adenine–thymine pair and three are involved in the cytosine–guanine pairing. The shapes of the molecules are such that these are the only pairings that allow the formation of hydrogen bonds, which explains the pairing of the bases attached to the two strands in the DNA molecule, as discussed earlier.

These hydrogen bonds are therefore responsible for the connections between the bases attached to the two backbones of the DNA molecule. The A–T and C–G pairings are the only ones that allow hydrogen bonds to form, which ensures that the sequence of bases associated with one backbone is fully determined by that of the other. The fact that hydrogen bonds are much weaker than conventional chemical bonds means that when the two backbones split apart in the process of reproduction or protein manufacture, the separation occurs by breaking the connections between the bases rather than the much stronger chemical bonds that attach the bases to the backbone.

Figure 4.6 shows the structure of the DNA backbone with all the atoms lying in the same plane. This provides a reasonably clear illustration of the way the atoms are connected together but it does not correctly capture the three-dimensional structure of the DNA molecule. When a carbon atom bonds to four other atoms (for example to four hydrogen atoms to form methane) it does so in a symmetric tetrahedral configuration. These bonds force the DNA backbone to twist, creating a helix. Of course, both of the backbones have this form, since their bases are linked together by hydrogen bonds. This results in the well-known 'double helix' of DNA, which was famously identified by James Watson and Francis Crick, using data obtained by Rosalind Franklin, who directed X-rays at a DNA crystal in the early 1950s.

The DNA found in human beings contains over one billion base pairs; if it had the form of a straight double helix, it would be about ten metres long but the DNA is contained in the cell nucleus, which is about 10^{-5} metres across. The relatively huge DNA molecule is packed into the nucleus by wrapping it around itself and some of the protein molecules, forming a series of approximately spherical sections that are connected by short strands of DNA chain. Each of these sections is known as a *chromosome*. Human DNA is divided into twenty-three chromosomes, each of which is about 10^{-6} metres in size. Two copies of all these chromosomes are contained in the nucleus of every cell in the body.

As discussed earlier, protein molecules consist of sequences of amino acids. The chemical structure of an amino acid is illustrated in Figure 4.8(a). It is composed of two carbon, one nitrogen, and two oxygen atoms, bonded together along with a 'side chain' (represented by the symbol R). There are twenty different types of amino acid, each identified by the chemical composition of its side chain. These range from a single hydrogen atom (in the amino acid glycine) to more complex atomic arrangements of carbon, nitrogen, and oxygen atoms, along with hydrogen and, in one or two cases, sulphur.

Amino acids can be connected end to end by removing the OH group from the right-hand end of one along with one of the hydrogens from the left-hand end of the other (these atoms combine to form a water molecule), which allows a C–N bond to form, connecting the two. This is illustrated in Figure 4.8(b), which shows a chain formed from three amino acids connected in this way. This process can continue again and again, leading to the long chains, containing anything between one thousand and one million amino acids, which form protein molecules. The globular shape of the protein molecule and the active area on its surface both result from the formation of hydrogen bonds and other connections between different parts of the chain.

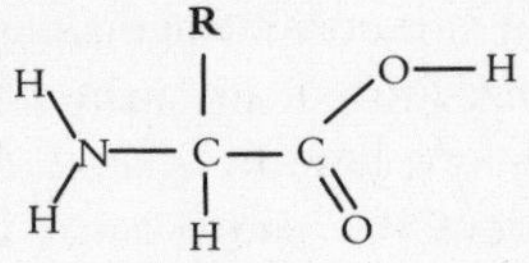

(a)

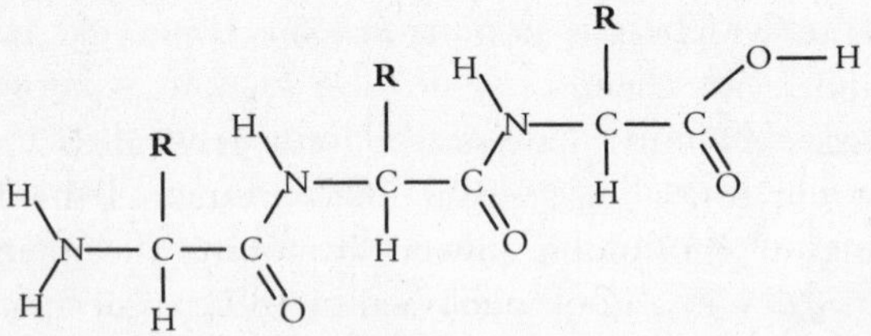

(b)

Figure 4.8 The structure of an amino–acid molecule (a) and how amino acids combine to form the chains found in proteins (b). Amino acids differ in their side chains, which are indicated by the symbol R in the diagrams. (Alastair Rae)

The evolution of life

The biochemistry involved in living organisms is almost unbelievably complex, yet this complexity is a natural consequence of the fundamental physical laws that control the behaviour of the constituent atoms of living things. The Schrödinger equation determines the motion of the electrons in the atoms and the possibility of chemical bonding emerges as a direct consequence. The particular nature of the bonding of carbon leads to the formation of the long chains that form essential parts of the molecules of both DNA and proteins. The role of DNA as a carrier of information supervenes on its chemical structure, as does its ability to use this in the form of a code that determines the properties of the proteins that constitute the cells of a living organism.

One big question, however, is still unanswered. Although this argument demonstrates that living matter is governed by the basic laws of physics, it says nothing about how this immensely complex system ever came into being. A fully reductionist explanation also needs to account for this in terms of these same laws. It was this problem that was considered to be the most challenging when the scientific study of life was being developed in the nineteenth century.

Before a scientific understanding was developed, the question of the origin of life was often treated as theological, requiring a miraculous answer. Famously, the theologian William Paley, in a book published in 1802, argued that anything as complex as life must have been the work of a designer. As an analogy, he suggested that if one were to come across a mechanical pocket watch lying on the ground, one would naturally conclude that it must have been designed and built by a skilled watchmaker. An important point is that there is no suggestion that the operation of the watch itself results from anything more than the known laws of physics, including Newton's laws of motion. It is the fact that it exists as a complex object that led Paley to conclude that it could not have come into being spontaneously but must have been designed by someone who had prior knowledge and understanding of what they were intending to produce.

Even before the internal structure and mechanism of living things had been discovered, it was well understood that living creatures, particularly humans and other animals – with hearts, lungs, means of locomotion, and so on – are a lot more complex than a watch, so Paley's conclusion that there must be a designer of living creatures seemed natural and obvious. As a theologian, he naturally assumed that the designer must be God and his argument is an example of the 'argument from design' for the existence of God. Paley extended his argument to claim that the fact that nature appeared harmonious, with every creature performing some function, is evidence that nature must have been designed by a beneficent God for the benefit of His creation.

One person who read Paley's book and was originally convinced by the watchmaker argument was the naturalist Charles Darwin. Charles Robert Darwin was born in Shrewsbury, England, in 1809. After abandoning a course in medicine, he studied for a BA at Cambridge, with a focus on both nature and theology. After graduation, he decided to undertake an in-depth investigation of the natural world, with the original intention of using this to further his understanding of the nature of God. He was offered and accepted a place on HMS *Beagle*, which was setting out on an expedition to chart the coastline of South America. During this journey, which began in 1831 and lasted for five years, Darwin studied the local flora and fauna he encountered extensively and in depth.

On his return, Darwin continued his studies, based on the notes he made during the voyage as well as the results of further observations and experiments he carried out afterwards. He found it increasingly hard to reconcile his observations with the idea of a beneficent creator because of the apparent lack of moral purpose in the alleged design, which was exemplified in the apparent cruelty that is often displayed in nature. He also became increasingly convinced that the contemporary form of the natural world, including the species of plants and animals, was not immutable but had developed over a long period of time by a process of evolution based on what is known as *natural selection*. The fundamental idea behind this is that as one generation succeeds another, living beings evolve so as to have the best chance of surviving in an often hostile environment. Whenever offspring are born, they are all likely to be a little different in some way from their parents and each other. Some will be more fitted to survive attacks from disease, food shortages, and predation by other creatures or anything else that threatens them. These are therefore most likely to survive and pass their advantages down to their offspring. Over many generations, these changes accumulate and later descendants can be very different from their original ancestors.

This idea has been studied and tested over a wide range of natural phenomena. Species have evolved in this way over millions, indeed billions, of years but significant changes can also be observed over a much shorter interval. In the eighteenth century, moths in the midlands of England were predominately white in colour. After the industrial revolution began, the environment was heavily polluted by the soot created by the burning of coal, which meant that white moths stood out against a black background and became very vulnerable to predators. Darker-coloured moths therefore had an advantage and quite soon, the standard colour of the species became black. Following the UK's Clean Air Act, around the middle of the twentieth century, the environment changed again and the population of moths responded by reverting to being predominately white.

The progress of evolution over much longer timescales has been studied using the fossil record, which dates back over millions of years. This has led to quite a detailed understanding of how life has evolved since its inception. There are still many questions that are not fully answered but there are no observations that would falsify the principle of evolution by natural selection from primitive single-celled creatures up to the present-day living world with its vast variety and complexity.

In developing his theory, Darwin made important assumptions about the nature of heredity, which he derived from his observations. One of these is that most characteristics of living creatures are transmitted from parents to their offspring, while the second is that this transmission also involves some random variations. At the time of Darwin's work, little if anything was known about the mechanisms underlying either of these. In the late nineteenth century, work done by Gregor Johann Mendel, a German–Czech Augustinian monk and scientist, on the nature of inheritance in plants became more widely known. Mendel studied the inheritance patterns of certain properties of pea plants and found that these could in many cases be explained through simple rules and ratios. This gave rise to the science of genetics and the idea of the gene as the unit of inheritance. What a gene actually consists of remained unknown until the chemical nature of cells and, in particular, the structure of DNA, became understood in the twentieth century. The results described in the earlier parts of this chapter can therefore be used to explain what is going on at the molecular level when Darwinian evolution takes place.

Reproduction

A universal feature of living things and a property essential to the process of evolution is their ability to reproduce. To understand this, we have to know a little more about what happens when an embryo, and eventually an offspring, develops from the fusion of an egg with a sperm. This process is amazingly intricate and complex and any attempt at a full description would lie well beyond the scope of a book like this but we can usefully note some of the main stages. A fertilized egg begins as a single cell containing DNA and proteins, much the same as any other. In the case of human beings, the DNA nucleus contains twenty-three chromosomes that originated in the female egg before fertilization and a further twenty-three that originated in the male sperm. This DNA will eventually be copied into the nuclei of all the cells in the resulting embryo and, via the newborn child, into the mature adult. Moreover, every one of these DNA molecules contains all the information required to generate copies of all the proteins and other components in any and all of the cells in the human body. Following fertilization, the egg cell divides in the way described earlier until a spherical structure, composed of about thirty cells inside a hollow sphere composed of other cells, is produced. The cells comprising this outer coating eventually create the placenta and umbilical cord, while the internal cells develop into the actual person. The process whereby this takes place involves a complex network of interactions between the genes and the proteins that ensure that the right proteins are manufactured at the right place and time for the development to proceed correctly.

During the development of the embryo, some cells develop to form the genetic material or *germ cells* – eggs in the case of the female and the sperm carried by the male – that can be used to produce later generations. An important stage in the development of the egg cells is known as *meiosis*; this involves the future egg cell undergoing two divisions where the DNA is reproduced only once, so that four copies are produced, each of which contains only one copy of the DNA molecule. A process of shuffling in which each chromosome acquires a mixture of the DNA produced by both the original parents also takes place during meiosis. As a result, each of the resulting four cells forms an egg that contains DNA from both the original parents but is still different from that originating in either. A large number of eggs are formed in this way and these are stored in the body in such a way that their DNA remains unaltered throughout the life of the organism. Sperm in the male is also generated and preserved during the embryo's developments but meiosis does not occur in this case.

The above process has both the features required for Darwinian evolution to take place: the genes contained in the eggs and sperm are inherited from both parents, so many of their characteristics are carried forward to the later generation. The shuffling ensures that the offspring are not exact copies of their parents but contain variations, which may make them more or less able to survive in their adult environment. Another source of variation is the occurrence of *mutations*, which consist of changes to the sequence of the bases in the DNA molecule. Mutations can affect significant sections of the DNA chain or only one of the bases at a particular site. They can occur in a number of ways, including mistakes made in copying the DNA during replication (though these are very rare), the result of the action of various chemicals which can enter the cell through the bloodstream, and the effect of radiation, including that associated with ultraviolet light. If – and only if – a mutation affects a germ cell can it change the form of the DNA inherited by subsequent generations and so play a part in the organism's evolution. This is why the inheritance of acquired characteristics does not feature in evolution: the germ cells are isolated from the rest of the body, so changes in body cells acquired during life, which may well affect their DNA, cannot be passed down to later generations. The dominant colour of moths changed from white to black during the industrial revolution not because pollution had made them dirty and they passed this feature on to their offspring but because in succeeding generations, moths that had acquired a darker colour from birth as a result of natural variation were better able to survive in the polluted environment.

Since Darwin's time, huge progress has been made in the study of evolution. Detailed understanding has been obtained of how various complex features of plants and animals (the eye, for example) have evolved from very simple organisms. The developmental biologist, Lewis Wolpert, has claimed that 'once cells evolved, it was basically downhill all the way to we [sic] humans'. A much harder question to address is where the first cells came from. Experiments were done in the 1950s, in which electrical discharges were passed through a mixture of gases containing methane, ammonia, and hydrogen, believed to be the composition of the atmosphere in primeval times. The result was that several amino acids were manufactured. How these came together to create the first proteins and how DNA and the other components of living cells appeared is still an open question and the subject of research. There is some evidence that the first self-replicating molecules may have evolved around hydrothermal vents in the deep ocean, where hot volcanic gases discharge into the sea. The search for evidence of primeval life in other parts of the universe may eventually provide information about how likely or unlikely it is for life to originate.

Many of the features exhibited by living organisms, including how a human being is produced from a fertilized cell, invite the word 'miraculous' – except that this would imply that something more than the laws of physics and chemistry are involved. The existence of life is an extreme example of the opposite: it is a prime instance of complexity emerging from and supervening on its component parts, which are subject only to the fundamental laws of science. The same physical principles that produce the solidity of ice and the liquidity of water also account for the emergence of the living cell from its component atoms and molecules and the creation of human beings with all their faculties and potential.

5

Neurons and the brain

In common with all mammals and many other creatures, human beings possess what is known as a *central nervous system*. This controls much of the body's activities by obtaining and processing information about its present state and sending signals to initiate and sustain actions. Moving a finger involves a decision in the brain, followed by the sending of a signal to the finger, instructing the relevant muscles to stretch or contract. All this information processing and signalling involves special cells, known as nerve cells or *neurons*.

Like other cells, a nerve cell contains a nucleus, where DNA is stored, and cytoplasm, where proteins are manufactured, but there are two important properties that only nerve cells possess. The first relates to their structure. Figure 5.1 shows a pictorial representation of a typical nerve cell. A long thin filament, known as a nerve fibre or *axon*, emerges from the cytoplasm. This is surrounded by a tough myelin sheath and is divided into sections by narrow necks, known as the *nodes of Ranvier*. Short branching sections, called *dendrites*, emerge from the cell body at the far end of the axon. The axon connects through the dendrites to other nerve cells or to muscles. These can be a long way away – over a metre in the case of a neuron connecting the brain to a muscle in the leg or foot.

The second property that is unique to nerve cells is their ability to transmit electrical signals along the axon. An essential requirement for this is a mechanism for the transport of electrical charge.

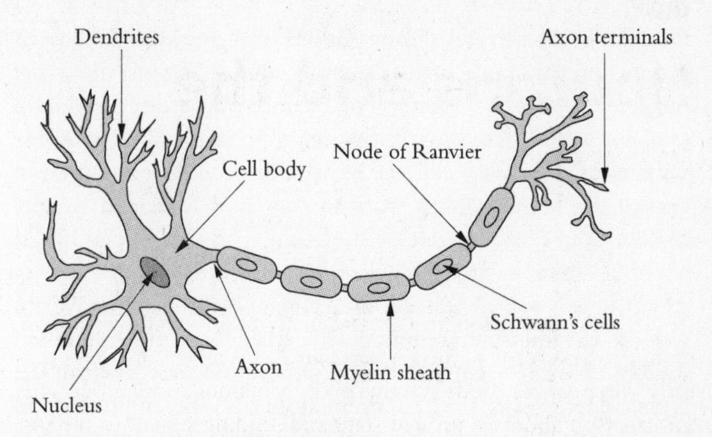

Figure 5.1 A nerve cell. (David Berger)

As mentioned in Chapter 2, metals such as copper contain 'free' electrons that are not bound to individual atoms, so they can move through the whole metal, carrying their electrical charge with them. One example of how this property is used to transmit information is a conversation conducted using two telephones connected by a copper cable. Each time one of the participants speaks, the sound produced is converted into an electrical pattern, which is reproduced in the motion of the electrons in the wire close to the transmitting phone. This sets up an electric field that induces a similar distortion in neighbouring electrons, which then transmit the signal to their neighbours. This continues until the signal reaches the receiver, where it is converted back into sound. The signal is transmitted between the electrons rather than being carried by a group of electrons moving from one end of the wire to the other. This means that the signal travels along the copper wire at close to the speed of light – much faster than the electrons could possibly move.

Similar principles govern the transmission of signals along nerve cells. In this case, however, the current is transmitted by ions, rather than free electrons. An ion is an atom that has lost or gained one or more of its electrons so that it possesses a net positive or negative charge. For example, an atom of sodium contains eleven electrons: the first ten of these fill all the available states in the lowest two energy levels, while the eleventh electron occupies a higher-energy state in the third level and is only loosely bound to the rest of the atom. The chlorine atom, in contrast, contains only nine electrons, which means that one of its ten lowest energy states is unoccupied by an electron. When sodium and chlorine are brought together form the compound sodium chloride – common salt – an electron moves out of the higher-energy state in the sodium atom into the empty state in the chlorine atom, leaving the sodium with a net positive charge and the chlorine with a net negative charge. These two ions are attracted to each other, lowering the total energy of the compound and making it more stable. This is true for common salt in its solid form and also when it is dissolved in water, when the resulting solution contains a mixture of mobile positive sodium and negative chlorine ions.

How does this relate to the operation of nerve cells? The fluid surrounding a nerve cell contains a mixture of ions, including those of sodium. The cell itself contains protein machines whose function is to pump sodium ions out of it. When the sodium ions leave the cell, a net negative charge builds up within the cell membrane. This acts to oppose further outflow of ions and a balance is established. The cell wall also contains a number of pores. These remain closed until they are stimulated by an electrical signal, which causes them to open, allowing positive sodium ions to flood into the nerve cell, neutralizing the negative charge and setting up an electrical pulse, known as an *action potential*.

This action potential propagates along a nerve fibre in much the same way as an electrical pulse travels along a copper cable. In a long nerve fibre, the signal weakens considerably as it moves along but at the nodes of Ranvier (see Figure 5.1) there are further sets of pores that open when the electrical pulse arrives, triggering a fresh action potential that restores the signal to its original strength. This is analogous to the presence of 'repeater stations' at points along communication cables, where signals are boosted to counter their inevitable attenuation. Although greatly simplified, this is the fundamental mechanism whereby information is transmitted electrically along a nerve cell. As with electrons in metals, there is no transport of ions along the length of the fibre; the signal is transmitted between the ions. As a result, signals travel along an axon quite quickly – at tens of metres per second – though this is still considerably less than the speed of an electrical signal travelling through a metal.

In the *motor neurons*, a bundle of nerve fibres terminates in a muscle. An action, such as raising a finger, is initiated in the brain, which creates action potentials in the relevant nerve fibres; these are transmitted along the nerve until they reach the finger, where they cause a series of chemical reactions to occur that result in the relevant muscles contracting and expanding and the finger being raised.

The brain not only sends signals out to the body, it also takes information in. All communication from sense organs to the brain occurs via the sending of electrical signals along *sensory neurons*. For example, special light-sensitive cells in the retina of the eye set up action potentials in the fibres of the optic nerve, which transmit signals to the brain. Similarly, sound waves cause the small bones in our inner ear to vibrate and stimulate special hair cells to generate action potentials which are transmitted along nerve fibres to the brain.

We can see therefore how physics and chemistry underlie biological action. The lifting of a finger and the sensing of visual information about the environment are both examples of higher-level phenomena that emerge from lower-level physical processes. Electrical signals move to and from the brain along nerve cells following the basic laws of physics and chemistry. Reductionism remains unchallenged.

There is, however, an 'elephant in the room': the brain itself. So far, I have said that the brain is capable of initiating actions by creating action potentials and processing information that reaches it from sense organs and so on. In other words, it has the capacity to think! For reductionism to apply fully to human beings, the principle has to underlie the whole process of thinking and feeling as well as acting. Indeed, this is probably the most radical and controversial step in the reductionist story and will be the subject of the rest of this chapter and the next.

The brain

The human brain has been described as the most complex object known in the universe – more complex than, say, an astronomical black hole or a molecule of DNA. How the brain works is a very active field of research but in recent years considerable progress has been made in unlocking some aspects of its basic structure and operation.

The general shape and structure of the human brain is shown in Figure 5.2. The brain is connected to the *spinal column*, the bundle of nerves within the spine that connects to the rest of the body. Sitting right on top of the spinal column is a section of the brain known as the *brain stem*, which includes the *medulla oblongata*, whose function is to control the automatic operation of the heart and lungs, regulating unconscious processes such as breathing, heart rate, and blood pressure. Alongside it, in the back of the skull, lies the *cerebellum*, which is responsible for co-ordinating the transmission of action potentials to the muscles. The main body of the human brain is called the *cerebrum* and the outer part of this is the *cerebral cortex*.

The cerebral cortex consists of four main parts, the *lobes*, which lie respectively at the back, top, sides, and the front of the brain. The rear or *occipital* lobe is connected to the optic nerves,

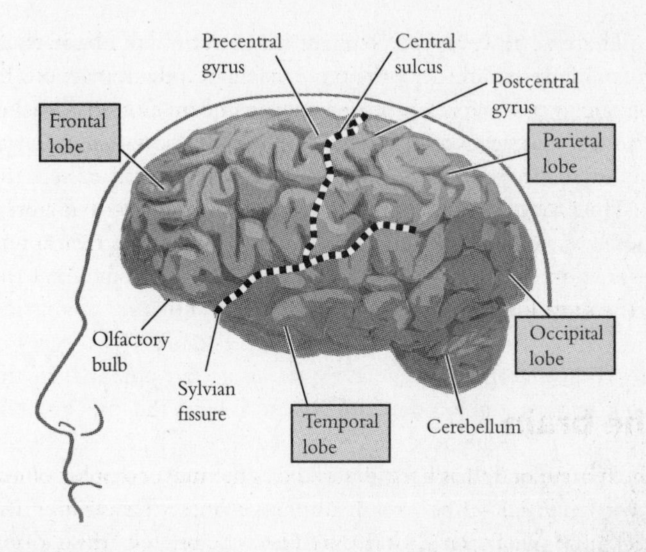

Figure 5.2 The human brain, seen from the left-hand side. (David Berger)

which emerge from the eyes. The top or *parietal* lobe and the side or *temporal* lobes control the other senses and are involved in the processing of language. Conscious thinking, understanding, and deciding take place in the *frontal* lobe. The cortex is also divided into right and left *hemispheres* by a fissure, so that these are approximately mirror images. Each hemisphere interacts primarily with one side of the body – the left hemisphere communicating with the right side and the right hemisphere with the left.

Like all living matter, the brain is composed of cells of various types but the most important are, of course, the neurons. There are somewhere near one hundred billion neurons in a human brain. While some are motor neurons and others are sensory neurons, the vast majority – the *interneurons* – are connected together. These may be connected to others that are nearby and to some that are some distance away across the brain. A large number of axons – typically several thousand – that originate on other neurons terminate on a typical neuron, which may also be connected to other groups of neurons via dendrites. It is within this vast, complex network that the phenomenon of thinking is believed to take place.

The point where an axon or dendrite connects to a neuron is known as a *synapse.* When an electrical signal, in the form of an action potential, reaches a synapse, it cannot directly cross the junction into the other neuron. Instead, it initiates the emission of a chemical *neurotransmitter.* There are essentially two types of neurotransmitter: one acts to excite an action potential in the target neuron and another operates to inhibit this process. The response of the neuron is determined by the balance between these two effects. If at a particular time, the total of the excitatory signals received from other neurons is greater than that of the inhibitory signals, an action potential is generated (the neuron 'fires') and a signal is transmitted along the axon to the next group of neurons. If not, the neuron does not fire and no signal is propagated. Whether or not a particular synapse fires at any given time depends on the total of the action potentials that have arrived at the transmitting neuron from all the other neurons it is connected to. Given the huge number of synapses associated with the massively complex network of connections between the neurons, this simple rule enables a huge number of possible synapse-firing patterns to occur.

When a neurotransmitter crosses from one neuron to another, further chemical processes take place that result in permanent changes in the neurons and, in particular, the strength of the synapses. These are consolidated if the pattern of neural firing is repeated. This is the basis of memory: a signal from a sense organ reaches the brain along a sensory neuron and triggers a particular pattern of neural firing, which persists and forms a record, or memory, of the original sensation. This process helps to create stronger connections between some neurons, when the synaptic path is used frequently in response to certain stimuli, and weaker connections between others. This property of the brain is known as *synaptic plasticity*, because the transmission of electrical signals between the interneurons changes the number of chemical neurotransmitters that each neuron releases and therefore the way in which the neurons respond. Such changes can be fleeting, in the case of short-term memory, or effectively permanent, when long-term memories are recorded.

Plasticity is particularly strong in babies and young children, as they acquire essential social skills such as language. Indeed, in infancy, many of the inter-neural connections are still being formed, so the formation of basic memories involves both the construction of part of the neural network and the establishment of firing patterns. This is believed to be what happens in the psychological process of *imprinting*, in which an individual rapidly learns about a particular new stimulus. A famous example of this involved the Austrian biologist Konrad Lorenz, who showed that newly hatched goslings would follow the first moving thing they saw, normally the mother goose but potentially any moving object – including Lorenz himself!

Looking inside the brain

Although there is a huge amount still to be learnt, scientists understand a great deal about what happens in the brain when various sensory and thought processes take place. Initially, this knowledge was confined to what could be gleaned from research on animals and on patients who had suffered brain damage. Since about 1990, however, it has been possible to obtain some information directly, by scanning the brain using Functional Magnetic Resonance Imaging (fMRI), a technique pioneered by the Japanese researcher Seiji Ogawa while he was working at AT&T Bell Laboratories.

fMRI is a development of MRI (Magnetic Resonance Imaging) which works by detecting the concentration of hydrogen atoms in an object. One of the properties of the protons that form the nuclei of hydrogen atoms is that they carry a small magnetic moment – that is, they are similar to very small compass needles. If a large magnetic field is applied to an object containing hydrogen, these little magnets line up parallel to it, in the same way as the Earth's magnetic field causes a compass needle to point north. This alignment corresponds to a lowering of energy in the system. If the magnets lined up in a direction opposite to the field, their energies would have been raised rather than lowered, so they would be in an excited state. We have seen how radiation of the correct frequency may cause a transition from a low- to a high-energy state, so if a further small field is applied that is vibrating at the correct 'resonant' frequency, the nuclei will be excited into this higher-energy state. Removing this oscillating field causes the nuclei to return to their original, lower-energy state, emitting radiation whose intensity is proportional to the concentration of hydrogen in the region being examined. It is possible to identify the radiation coming from a particular part of the sample and so measure the local concentration of hydrogen. By changing the position of this focus, an image of the object is created showing the concentration of hydrogen at different places.

MRI is widely used in medicine to investigate the body without using invasive procedures such as surgery. Unlike X-rays, which are mainly sensitive to the presence of hard tissue such as bone, MRI is able to create an image of the soft tissues and expose abnormalities such as tumours. In fMRI the same principles apply. The detected signal comes from the hydrogen nuclei contained in the water molecules that form part of the blood. As mentioned earlier, haemoglobin transports oxygen along the blood stream to the cells of the body, where it sustains the conversion of food to energy. The oxygen atom carries a magnetic moment, which produces a magnetic field that acts on the hydrogen nuclei in the surrounding water to produce a small but detectable change in the frequency of the MRI signal. This provides a measure of the amount of oxygen being carried by the haemoglobin molecules. When neurons fire, they consume energy, which must be obtained from the blood flowing nearby, so neural firing results in increased flow of oxygenated blood to that area of the brain and a consequent change in the signal detected at that locality. In this way, an fMRI scan can be used to create images of where neuronal activity is taking place in the brain.

This property is exploited by researchers to see what parts of the brain are firing when a person undergoes different experiences. Someone undergoing an fMRI scan might be subjected to a sensory stimulus, asked to consider a specific topic, or remember a specific emotional experience. While the subject is doing this, blood flow in the brain is observed. As a result, scientists have determined that activity in certain parts of the brain is involved in certain types of thinking. The method has limits: in particular, its spatial resolution is limited to regions around one millimetre in size or larger. This means that it cannot be used to examine the action of individual neurons but only the average properties of a brain region of that size, which generally contains a few million neurons.

Some information on what is going on at the level of a single neuron has been obtained through experiments on animals (monkeys, in particular) that have had electrodes carefully and humanely inserted into the brain. Similar measurements have been made on human subjects while their brains were being operated on for other reasons but even with all this information, scientists are still far from understanding the operation of the brain at a cell-by-cell level.

Consciousness

Consciousness is a remarkable property of the human brain. Although we all experience this, it is a difficult thing to define.

It involves the idea of the 'self' as the person who experiences the world and makes decisions on what to do. On the other hand, many of the actions we take do not appear to involve conscious thought and can be completely automatic – our breathing and the beating of our hearts are obvious examples.

To explore further the distinction between conscious and unconscious actions, imagine I am walking along a road and it starts raining. I do not have a coat or an umbrella but I see that there is a shelter a short way off, so I run there and wait for the rain to pass. One description of this scenario is that, once I became aware of the rain, I made a conscious decision to seek shelter and then did. An alternative account is that:

1. The physical processes governing the weather caused it to rain in my vicinity.
2. The impact of the rain on my face and other exposed areas of my body caused electrical signals to pass along my sensory nerves to my cerebral cortex.
3. The reception of these signals in my cerebral cortex caused motor neurons elsewhere in the brain to send electrical signals along the nerves that control the motion of my legs.
4. As a result, my legs moved and carried my body to the shelter.

In this, my brain acts much as a computer might; indeed, it is well within our present technology to build a computer-controlled robot that would mimic these behaviours.

What is the difference between the two accounts of my behaviour? The robotic description contains no reference to my making a 'conscious decision' and so gives no explanation of why, for example, step 3 occurs. What is it about the signals received from my exposed skin that causes my motor neurons to send the message to my legs? Obviously, getting wet in the rain is an unpleasant experience, which triggers a pattern of neural firing associated with similar previous unpleasant experiences recorded in my memory. This firing activates another pattern of neural firing that informs me that the best way to avoid this unpleasantness is to activate the motor neurons required for me to run for shelter. The robot also reacts to the rain by running for shelter but this is a programmed response that occurs even though the robot presumably has no concept of unpleasantness.

A surprising number of the actions we perform every day can be described as robotic in the sense that they happen without conscious thought. The blink of an eye to protect itself from being damaged by a rapidly approaching object takes place almost instantaneously and much more quickly than it would if the information were sent to the brain to be processed afresh each time. Even running for shelter from the rain could become automatic if I did it often enough: I might say that I had reached the shelter 'without thinking about it'.

Another example of robot-like behaviour occurs when a sport is played by practiced individuals. Consider two people engaged in a game of tennis. You might think that when a player hits a ball, the opponent's eyes observe the resulting motion of the ball and his or her brain consciously plans and executes a return shot. But this is an inadequate description of the game and how the brain works. I am certainly no tennis star but I know that the movement of the hand carrying my racket is consciously planned only in the most general terms. What actually happens is that my eye observes the motion of the ball and passes a signal to the brain that initiates an unconscious response resulting in my hand placing the racket in (hopefully) the right place to execute a winning return. If I look at where my hand is positioned afterwards, I can often be quite surprized at the result.

A less well-known example of robotic behaviour is *blindsight*. Some people who have experienced damage to the part of the brain that controls vision are blind in the sense that although they have no conscious awareness of a stimulus such as a flashing light, they may still be able to point in the direction the light came from.

It is believed that this process relies on an additional connection between the retina of the eye and a part of the brain that is not involved in conscious thought but which can still initiate motor actions such as pointing.

What then is the purpose of consciousness? Why are human beings not just robots? Before trying to answer this, we should try again to define what consciousness is. This is surprisingly difficult and controversial, given that consciousness is something we all possess and experience, but it is possible to identify some of its properties. One such is our ability to describe, say, a visual experience to someone else. This means that the experience has been understood and that you have been able to translate it into a symbolic representation – that is, verbal language. Memory is an essential prerequisite for describing a previous experience; if we could not remember anything for even a short length of time, we would not be able to describe it. Indeed, if we had no memory at all (short, medium or long term) conscious awareness would not be possible.

Another important property of consciousness is awareness of the self or *self-consciousness* – having a sense of identity, knowing that one exists as a different entity from other people. A simple manifestation of this is the ability to recognize your image in a mirror as you. This has been tested by marking the noses of children and then asking them to look at their reflection. Children younger than about eighteen months usually reach towards the mirror – apparently seeing the image as another child. Older children generally touch their own nose, showing that they recognize the image as a representation of themselves. It used to be believed that self-consciousness is a uniquely human attribute and that animals react to their mirror image the way an infant does. There is now, however, quite strong evidence that self-consciousness is also possessed by other animal species, such as apes, and even some elephants. In one experiment, an elephant whose face had been marked responded to seeing its reflection by trying to wipe the mark off its own face. Not all elephants or apes are able to pass this 'mirror test' but some apparently do.

This brief outline of some of the properties of consciousness offers some clues as to why the evolution of consciousness gave an advantage to those who possess it. Without conscious memory, planning future actions on the basis of past experience would be impossible, as would communication using language. Without the ability for conscious thought, it is hard to see how the human ability to make tools and build sophisticated shelters could have evolved.

Earlier in this chapter we saw how the activity in the brain consists of patterns of electrical excitations of neurons. How something as powerful and mysterious as conscious thought emerges from a complex pattern of electrical signals remains a major challenge to our scientific understanding. There has been considerable recent research aimed at identifying which brain patterns are identified with consciousness. One such set of experiments exploits a phenomenon known as *binocular rivalry*. Two different images are directed at a subject in such a way that he or she sees one of them in the left eye and the other in the right. The subject consciously sees each image alternately, even though both are impinging on the retina simultaneously and both are connected via the optic nerves to the cerebral cortex. The aim of the research is to detect where, in the passage of signals from the eye to the brain, the choice is made of which image to see. Much of this work has been done by inserting electrical probes in different parts of the brains of monkeys to observe the operation of individual neurons. The monkeys have been previously trained to press one of two buttons, depending on which image they are aware of. At least one similar test has been made on a human patient who was undergoing a brain operation to treat epilepsy. In all cases it was found that the signal remained the same until it reached a region of the brain, the *auditory visual cortex*, in which different neurons fired depending on which image was being perceived. Research of this kind is beginning to identify what are known as *neural correlates of consciousness* (NCC). The aim is to identify in as much detail as possible which parts of the brain are involved in conscious perception and indeed, introspective thought.

Even if all NCCs are fully identified some day, it is not clear that this would produce a consensus on the nature of consciousness. Certainly, no such consensus currently exists, which means that the field is still open to speculation, particularly among philosophers who have proposed a number of theories of consciousness, not all of which accept reductionist principles. A full understanding of the nature of consciousness is still one of the great unsolved questions in philosophy and science but if we follow the line of thought we have applied earlier, we should assume that it is an emergent phenomenon that supervenes on the electrical activity of the brain, unless and until we encounter evidence that falsifies this. The philosopher Daniel Dennett, among others, argues strongly for this point of view. He describes the alternative as believing that there is a 'Cartesian theatre', in which the brain presents a 'performance' that is observed by a non-physical mind. Although we might think that this describes our conscious experience quite well, Dennett strongly argues that this is an illusion. Our brains generate this illusion but fundamentally it is just another way of describing the actions of the brain. Others have produced arguments that claim to prove that a reductionist explanation of consciousness is logically impossible. The next chapter explores consciousness further and considers some of the attempts that have been made to falsify the reductionist account.

6
Can we reduce the mind?

Earlier chapters have discussed a number of examples where complex higher-level properties emerge naturally from lower level substructures: water is liquid, ice is solid but both are composed of the same type of molecule obeying the same laws of physics. The properties of biological cells result from chemical reactions involving the DNA in the nucleus and the proteins in the cytoplasm, while brain functions emerge because the neurons in our central nervous system are connected by synapses that either fire or don't fire according to the electrical signals reaching them. Can the same reasoning be extended to the operation of the conscious mind? Shakespeare's Hamlet asked this question rather more poetically:

> What a piece of work is a man! How noble in reason, how infinite in faculty! In form and moving how express and admirable! In action how like an angel, in apprehension, how like a god! The beauty of the world. The paragon of animals. And yet, to me, what is this quintessence of dust?

A modern, more prosaic version of Hamlet's question might be 'can the capacities of human beings (man or woman!) to move, reason and understand, to perform acts of heroism, but also of great cruelty, to deploy complex rational arguments, but also behave irrationally, be reduced to a "quintessence" of electrical signals in a brain?'

Mind and body

The idea that conscious experience is something distinct from the physical states of the brain has a long historical tradition. In many cultures, the mind, or soul, is considered to be essentially different from the material body, including the brain. Indeed, most religions maintain that a human being has an immortal soul that survives the death of the body, to be resurrected in another (hopefully heavenly) existence or perhaps to be reincarnated in another body, human or animal.

The idea of the mind as an entity separate from the brain was extensively developed by René Descartes, the seventeenth-century French writer, who has been called the father of modern philosophy. Descartes contributed to a number of areas of thought, including mathematics, where he invented what we now call *Cartesian co-ordinates* (a way to describe the position of any object by three numbers: its distance forwards, sideways and upwards from a fixed point). He was also one of the first to formalize a philosophy of the mind's relationship to the body.

Following from his famous statement *cogito ergo sum* ('I think therefore I am'), Descartes proposed that our consciousness is all-important and resides in the mind, which he believed must be a separate, non-material entity. The mind controls our thinking and decision-making, while the brain operates like a machine following physical laws. This implied that the mind and the brain must interact in some way, since the former has to be able to get information from, and pass messages to, the latter. He suggested that this function was performed by the pineal gland, a small object in the brain that was later found to be responsible for the production of a hormone that helps regulate sleep patterns, among other things. Descartes made the suggestion partly because the physical function of the pineal gland was not then known and it was incorrectly believed that the organ was unique to humans. Because of his emphasis on the separateness of body and soul, these ideas are often described as *Cartesian dualism*.

At every stage where reductionism has been used previously, the principle of falsifiability has been applied. For example, the properties of ice and water are fully consistent with the laws governing the behaviour of their component atoms and molecules. In deciding whether the same reasoning can be applied to the relationship between the human mind and the brain, we should not be influenced by the fact that we experience the amazement and admiration expressed by Shakespeare. Given our general assumption that reductionism holds until it has been falsified, the only valid reason for rejecting the application of this principle to human consciousness would be if a contradiction could be established between our mental capabilities and the laws governing the lower-level operation of the brain. Few would believe that Descartes's arguments achieve this but there have been many attempts over the years to demonstrate that just such a contradiction exists.

Free will

The ability of human beings to make conscious decisions was discussed toward the end of the previous chapter. Consciousness leads to the idea that human beings possess *free will* – that is, the ability to choose freely between possible courses of action. From a dualistic point of view, free will is assumed to be a property of the mind; if it is something that supervenes on the physical brain, it is hard to see what it is that does the 'choosing' or in what sense the choice is 'free'. For some time now, however, scientists' investigations have thrown doubt on the existence of free will. In the 1970s, the American scientist Benjamin Libet carried out experiments into the neural activity of the brain as part of his research in the physiology department of the University of California in San Francisco. In some of these, he investigated what was going on in the brains of subjects when they made a conscious decision to carry out a simple action, such as pressing a button. He attached electrodes to his subjects' scalps, to measure the brain activity associated with a signal to the hand telling it to perform an action. He then asked his subjects to press a button that was connected to a light and the resulting brain signal was detected. Libet found that the signal occurred about half a second before the action of pushing the button began, indicating that it took this time for the message to be transmitted along the nerves from the brain to the hand.

In another set of experiments, Libet asked his subjects to look at a clock while making a conscious decision to push the button and to remember the time they made this decision. Surprisingly, he discovered that the button was pushed only about two-tenths of a second after the conscious decision to do so, which, given his earlier results, is three-tenths of a second *after* the signal to push had left the brain. Indeed, in other experiments, the delay between pushing the button and deciding to do so has been found to be as long as one second. It appears that what the subject thought was a conscious decision to act had already been made in her brain without her knowing it!

These results have spurred considerable controversy. On the one hand, it has been suggested that they mean that a conscious decision is capable of acting backwards in time to affect the brain (some three-tenths of a second in the past), while, on the other, it has been taken to imply that the idea of conscious choice is an illusion. If the former were the only explanation, reductionism would certainly be falsified. But this should only be considered if alternative explanations fail.

One key feature of Libet's experiments was that the subjects knew that they were going to initiate an action at some point in the next few seconds. Their 'free choice' related wholly to the particular time they did so. You can try this part of the experiment for yourself without the need for brain-monitoring equipment. First, plan to move your finger sometime during the next few seconds and then make a conscious decision actually to do so.

Do you know why you chose to move your finger at the particular time that you did? Or does the action seem to be at least partly involuntary? Could it be that your brain sent the signal to your finger before it made you conscious of this fact? It appears that when we make choices of this sort, both the conscious and the unconscious are involved.

Even if the unconscious part of the brain is largely responsible for taking quick, nearly instantaneous, actions, how do we make longer-term policy decisions, such as the one I made to embark on the project of researching and writing this book? The relevant factors that influenced me in this case included my previous experience in writing, my interest in particular questions over the years, the fact that I had time to devote to it now that I have retired from full-time academic work, the encouragement of others, and so on. My brain processed all these inputs and produced a mind-set that largely determined my course of action. Further support for this model of decision-making comes from the common practice of 'sleeping on it' before finalizing a decision. Sleep seems to give the unconscious mind time to process all the relevant information; this is channelled into our consciousness, which then makes the decision that we 'feel' is right.

The natural reaction of participants in the Libet experiments was to say that they could have decided to press the button at a different time from when they actually did – that this decision was an act of free will, not one in some way pre-determined by the unconscious parts of the brain. Reconciling free will with the apparent determinism of a material universe composed of atoms following set laws has exercised the minds of many thinkers over the ages. The Greek philosopher, Epicurus (341–270 BCE) believed that the universe was fundamentally a mechanical system. He tried to include the possibility of free will by postulating the idea of a 'swerve', meaning that atoms undergo small random deviations from their path as set by the laws of mechanics, and that this deviation could in some way be affected by the human soul. This idea re-emerged in the twentieth century, when it was realized that randomness was a feature of quantum physics. How the soul is supposed to affect Epicurus's swerve is not at all clear, however, and in the quantum case any such influence would only be possible if the expected randomness were capable of being affected by a non-physical entity. This would seem to be just as radical as saying that a non-physical mind could affect the brain directly, even if there were no randomness. In view of this, we can conclude that the idea of free will is an illusion, at least in contexts similar to that in Libet's experiments.

Even if our decisions are determined by the physical state of our brains, these are *our* brains. The decision about when to push a button or whether to write a book is made by us, in the sense that the physical states of our brains have been influenced by the sense data they have received and the conscious and unconscious thinking processes they have undertaken. In this sense any decision is freely made by our whole selves.

Mary and the colour red

What is the colour red? Could you identify it or describe it to someone who has never seen it? Could you explain the brain processes involved in being able to sense that something is red as opposed to being green, or colourless?

Human beings directly experience sensations such as hunger, pleasure, and pain and perceive properties such as colour, musical notes, or the taste of food. When I see red, I may know that I am sensing light whose wavelength lies within a particular range of values but this description seems to have very little to do with what I am *experiencing* – indeed, most people experience colours without knowing anything about the theory of colour vision. We are also able to group objects together based on general, abstract concepts. We can say that a collection of things is all red, even when each thing has a slightly different amount of 'redness'. Consider, for example, being presented with an object that is on the orange side of red or the red side of orange. In these situations, we make a decision about whether the object is red or orange. The experience of redness, along with other sensations, is an example of a *quale* (plural *qualia*) – a quality of an object that cannot be understood except through sensory experience. Imagine trying to explain the concept of redness to someone who is colour-blind.

Such reasoning has led some people to believe that qualia represent something beyond what can be comprehended by a physical human brain. In the 1980s, the Australian philosopher Frank Jackson created a thought experiment which was intended to show that a quality such as the colour red could not be fully understood, even if it were possible to describe all the brain processes involved when it is experienced. He imagined a character, Mary, who has had an extremely unusual upbringing. She has been thoroughly educated in all aspects of visual perception: she knows exactly what the eye and the visual cortex do and indeed how the whole brain behaves when observing coloured objects. In this thought experiment, we may imagine that her knowledge extends well beyond what is currently known. Mary has, however, spent all the days of her life so far in a house in which there is no colour, so that her direct sensory experiences have been limited to shades of grey.

Mary is allowed to leave her home, and is shown a red rose. In view of her limited experience but exhaustive knowledge, does this event produce a new sensation? Jackson said that it would and that it leads to this argument:

1. Before her release, Mary knew everything that could be known about the physical processes associated with seeing the colour red.

2. After her release, Mary has learned something new about what it means to see red.
3. Since she already knew everything physical about redness, the new information cannot be physical.
4. Hence our mental experiences must be more than can be represented by our physical brain.

This argument rests on the assumption that what Mary's conscious mind experiences when she sees the rose is different from anything she encountered while undergoing the learning process. If this is the case, however, there is still no need to postulate the existence of something beyond the physical brain. When Mary sees the rose, the physical signals that pass through her eyes, optic nerve and visual cortex and into her brain will be different from any that she has previously experienced, even though this included all there is to know about colour vision. The state of her brain after this experience will therefore be different from what it was previously. A new pattern of neural firing will have been triggered, from which the experience of redness will emerge and supervene on this state of the brain. Her knowledge has indeed increased, in the sense that she now knows what this experience is.

In other words, knowing about colour and experiencing colour are different processes, though both are performed by the brain. Thus, if the hypothetical Mary were put into an fMRI scanner so that we could see what was going on in her brain before and after her first encounter with the red rose, we would see a pattern of neural firing different from any of the patterns formed during her study of colour vision. Even if it is correct to describe this as 'non-physical' (statement 3 above), it is still perfectly possible that it supervenes on the physical brain and statement 4 does not necessarily follow from 3.

An alternative approach to this problem is to ask what complete knowledge of visual perception should consist of. If this is defined as including that obtained from experience, then for Mary to really know all about colour, she would actually have already induced the experience of redness into her brain (perhaps by some futuristic process involving surgery and/or drugs) and seeing a rose would tell her nothing new. Statement 2 would then be incorrect and the rest of the argument would fail. Jackson's argument seems to assume, without justification, that knowledge by anything other than textbook learning or something similar is non-physical.

Generalizing from this, the reductionist assumption that qualia, such as the experience of the colour red, supervene on the physical brain has not been falsified by this argument.

The Turing test

In recent years, great advances have been made towards understanding the operation of the brain but much more progress will be needed before the processes involved in an experience such as seeing a red rose are fully identified. Another way of exploring the relationship between the mind and the brain is to look for an alternative physical system that has properties similar to those of the brain and whose detailed operations can be studied, to see how its large-scale properties supervene on its internal structure. If this system could be induced to undergo experiences and adopt behaviour that is essentially the same as those found among human beings, we would have very strong grounds for believing that the human mind supervenes on the brain.

Much of the work in this area has concentrated on developing computer programs capable of exhibiting what is known as *artificial intelligence*, often abbreviated to AI. Could it be possible to write computer code that enables a computer to reason and think in a manner that is indistinguishable from a human mind? If so, we could use the operation of the computer to learn what is involved in the experience of reasoning and thinking. Conversely, however, if it could be shown that there is something that human beings do that is in principle impossible for any computer, this model of human thinking would be falsified. Many critics of AI research claim that there are indeed mental processes that will never be modelled computationally.

The first step towards understanding these criticisms is to ask what it is that computers actually do. Nowadays, computers are very powerful processors of information, so much so that they can do many things much better than most humans can. A quite basic computer can total a million numbers in a fraction of a second and with essentially zero probability of error, while a human would take many hours to perform the same task and would almost certainly make some mistakes along the way. Nevertheless, these sorts of skills are not usually regarded as being evidence of significant intelligence but only of the ability to perform a mechanical task. Using a computer as an aid to do arithmetic is rather like using a lever to move a large weight: neither is evidence of a significant level of understanding on the part of the device.

Beyond this, however, computers can also be programmed to produce graphical images, play music, evaluate the results of mathematical expressions and play games, such as chess, that are believed to require a high level of mental ability when played by us. When a computer defeats a chess grandmaster, this is certainly evidence of a skill that is similar or even superior to that possessed by many human brains. However, this in itself is not sufficient evidence to show that everything brains do can be simulated by a machine. To establish this, it would be better to have objective criteria to decide whether or not a computer program is operating in a manner indistinguishable from that of a human mind.

Scientists working in this area have developed such criteria, which are expressed in what is known as the *Turing test*. Alan Turing was a brilliant mathematician who was highly influential in the development of computer science. During World War II, he worked in Britain's code-breaking centre at Bletchley Park and for a time was head of Hut 8, the section responsible for German naval cryptanalysis. He developed an early version of a computer (known as the 'bombe') that played a vital role in decoding messages sent by the German Enigma coding machines. After the war, he committed suicide, following harassment by the authorities over his homosexuality; this was the subject of an apology by the UK prime minister, Gordon Brown, in 2009.

Turing thought deeply about the nature of computers well before they had been realized in practice. He foresaw the possibility that they could develop powers similar to human thought. He proposed a procedure that could be used to test a machine's ability to exhibit intelligent behaviour: a three-way conversation involving a computer and two human beings, all hidden from one another. (To make it fair to the computer, both people type their contribution to the conversation and everything is read on computer screens.) One of the humans is asked to assess which of the two streams of dialogue he is receiving is 'more human'; if he cannot tell the machine from the human, the machine is said to have passed the Turing test.

Despite some false dawns, however, computers still seem to be a long way from being able to trick humans into thinking that they too are human. But if a machine were developed that could pass the test, it would indeed provide strong evidence that our thinking processes supervene on the operation of the neurons in our brain in the same way as the higher-level properties of the computer emerge from the electrical behaviour of its components. This does not mean that there has to be an exact correspondence between the individual neurons and the computer components, only that the same principles of reduction and emergence can be applied. AI enthusiasts believe that it is only a matter of time before a machine capable of passing the Turing test is developed but there are others who believe that, even if this happens, it will never be possible for a computer to think in the way humans do. Other critics of this approach question whether passing the Turing test would be sufficient evidence that a computer had a mind like that of a human being.

The Chinese room

Among those who believe that even if a computer were developed that could pass the Turing test, it would still not be able to do everything that a human mind can, is the American philosopher, John Searle, who was Slusser Professor of Philosophy at the University of California, Berkeley. In a paper published in 1980, which subsequently featured in his BBC Reith lectures in 1984, he suggested a scenario in which a computer was able to converse in Chinese. If it could pass the Turing test, the Chinese-speaking computer would behave in a manner that is indistinguishable from that of an actual person communicating in Chinese via a keyboard and screen.

Imagine that, instead of the computer, we have a room containing a store of Chinese characters, a copy of the code previously used by the computer and an operator who understands what computer code does but has no knowledge of the Chinese language. The operator can use the room to converse with a human partner speaking Chinese:

1. A set of Chinese characters is passed into the room.
2. The operator looks at each character in turn and consults the computer code to discover how the computer would have processed it.
3. The operator follows exactly the same procedure as the computer did and is guided to select another set of Chinese characters from the storeroom.
4. These characters are passed back to the person outside the room.

In this way, a conversation is conducted that is identical to the one conducted with the computer before it was replaced by the room. (The process presumably takes place much more slowly but that is not relevant to the argument in principle.)

Searle emphasized that, because the operator knows no Chinese, she has no knowledge or understanding of the content of the conversation. Since the operator is doing exactly what the computer did, it follows that the original computer program could not have had any knowledge or understanding either. This means that no computer could ever truly understand and think in the same way as a human being. The suggested conclusion is that human qualities such as awareness and understanding are in principle beyond the ability of any computer, whether or not the computer finds a way to pass the Turing test.

There is no doubt that Searle's Chinese room presented a considerable challenge to AI researchers, especially those who believed that they were getting close to achieving their goal of programming a computer to pass the Turing test. They countered Searle's arguments in a number of ways – although none have convinced Searle to change his mind. One of the difficulties of this debate lies in defining what we mean by a term like 'understanding'. Suppose two people are conducting a conversation in their common native tongue. One person says to the other, 'I see the sun is shining today'. To understand this sentence, his conversational partner must call on her memories of similar situations: the sun is an object in the sky that shines (that is, emits light); 'today' means the time since she woke up and includes the present instant. She might also recall that on previous occasions when the sun was shining, this triggered a response in her nervous system and then in her brain, to which she has ascribed the word 'hot'. So she might reply, 'Yes and this means that the weather is hot'. Without such memories, the words she heard would have no meaning and she would be unable to conduct an understandable conversation.

Before a computer could have any hope of passing the Turing test, it would therefore need to have the capacity to call upon memories of past experiences. The instructions the computer followed would have to involve relating and correlating these experiences with language in a way that is at least analogous with the operation of a human mind. These memories might have been of real experiences or, more likely, would have been implanted in the computer's memory during its original programming. In any case, the existence of memories is essential to the ability to pass the Turing test.

The same would be true of a human being, of course. If a person suffered from illusions or from false memory syndrome, her replies in a conversation might make little or no sense. She might produce a nonsense answer to the statement about the weather: 'Yes and this means that the air makes a sweet sound'. In which case, she would also fail the Turing test.

It follows that 'understanding' is not just a property of some part of the brain (or a particular part of a computer); it is a description of a *process* that involves large portions of the brain acting together. Suppose we could look into the brain of someone who is able to speak Chinese and is holding a written conversation in this language. One of the actions of the brain would be to accept and record the Chinese symbols as they are read; another would be to pass instructions to the person's fingers on how to write out the reply. Neither of these parts of the brain would understand the meaning of the symbols. Understanding is a property of the whole brain, or at least the large part of it that is involved in thinking. When a human mind (or a computer) speaks Chinese, the understanding involved is a property of the whole system. In the same way, we should not expect the person in Searle's Chinese room, who acts as an automaton, to acquire understanding of the meaning contained in the Chinese characters. To pass the Turing test, it must be the whole room, rather than the operator, that speaks and understands Chinese. From this point of view, the Chinese room teaches us not to look for a 'centre of understanding' in the brain but to see understanding as a higher-level property of the whole mind.

We can illustrate this point further by another fanciful argument. Suppose we understand the operation of the brain well enough to know that, as a conversation proceeds, a particular neuron is always involved. It may fire or not fire, depending on the inputs it receives but this firing forms an essential part of the process. Imagine that we were able to remove this neuron from the brain and insert one wire that can detect the strength of the signal the neuron would have received, and another wire through which we could send impulses to other neurons that were connected to the now-absent neuron. As the conversation proceeds, imagine that a human engineer monitors the inputs to the circuitry that has been substituted for the neuron and sends pulses to other neurons whenever she detects that the neuron would have fired. Provided this is done quickly enough, the conversation will proceed exactly as if the original neuron had not been removed. The engineer, however, is aware only of the signals arriving and leaving this neuron and can certainly not use this information to obtain any significant knowledge or understanding of the actual conversation. Just like the operator in the Chinese room, the engineer is performing an automatic task. In contrast, understanding is a property of the whole mind.

The Penrose argument

The English mathematical physicist Roger Penrose shared the Wolf Prize for physics in 1988 with Stephen Hawking for contributions to our understanding of black holes and the universe. He also takes a close interest in the properties of the mind; in a series of books that began with *The Emperor's New Mind*, published the year after the Wolf Prize award, he claims that human beings must be capable of a deeper understanding than can be achieved by a machine such as a computer. In other words, thought cannot be completely reduced to computation. Penrose's argument relies on showing that, however powerful a computer we construct, there are always some questions that a human can answer but a computer cannot. These generally involve some form of self-reference, where a question is asked about some property of the question itself.

In this book I have tried to avoid technical details wherever possible but an understanding of Penrose's argument that human thinking is more than computation requires a basic knowledge of what a computer does. Typically, a computer is *programmed* to perform a defined set of operations or instructions in sequence. A very simple computer program, which is shown in Figure 6.1(a), is one often used to test a new computer system. It consists of only two instructions: the first results in the computer printing out the message 'Hello World' and the second stops the program. A slightly more sophisticated, three-instruction program is illustrated in Figure 6.1(b). After printing the message, the program moves on to the second instruction ('CONTINUE TO 3'), which transfers the computer to the next instruction ('REPEAT 2'), so setting up what is called an 'endless loop' between instructions 2 and 3.

These two simple examples illustrate a general feature of computer programs, which is the *halting* property. Either a program comes to a stop at some point in its operation or it continues indefinitely – until, that is, the computer is switched off. This raises the question of how we can tell what any particular program will do – will it halt or will it go on for ever? This halting problem might seem to be a rather technical question that should only be of interest to computer specialists but it actually leads quite directly to what is claimed to be a fundamental distinction between the operation of a computer and the workings of a conscious mind.

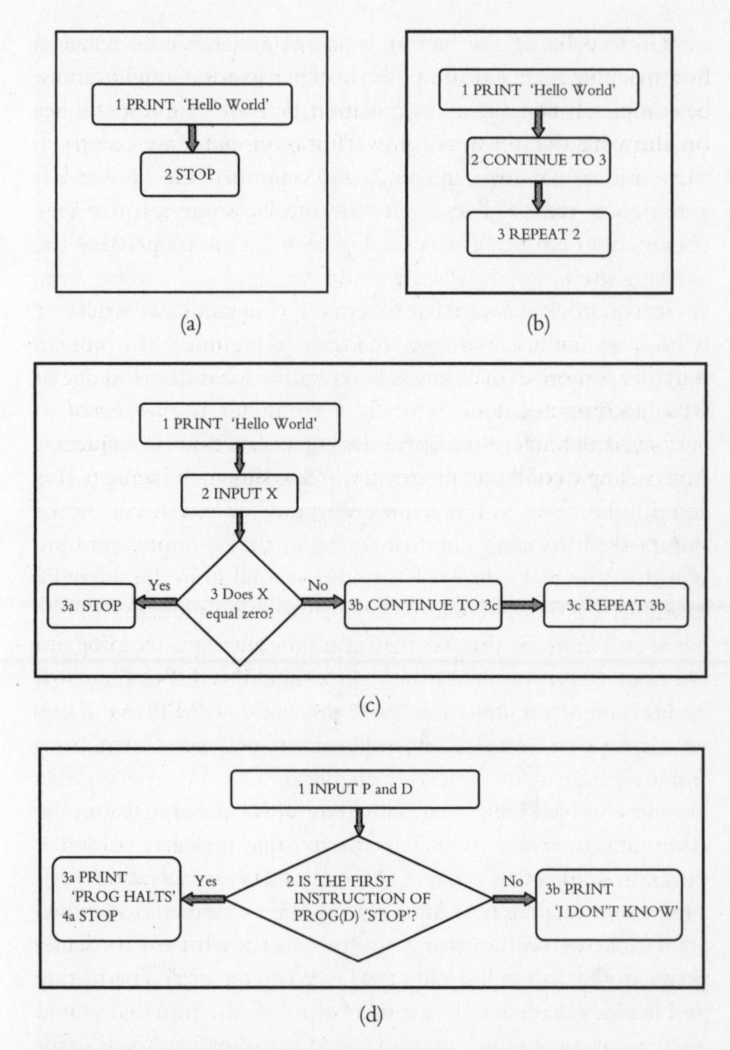

Figure 6.1 Four simple computer programs. Program (a) halts after printing out a simple message, while program (b) prints the same message and then starts an endless loop. Whether or not program (c) halts depends on the value of the parameter X that is read from the keyboard. Program (d) is a simple version of HALT. (Alastair Rae)

Understanding the halting problem requires knowledge of another important feature of computer programs, which is that they may contain 'branching' instructions. These mean that the program follows one or another set of procedures, depending on the value of some parameter, which may be entered into the program as data. For example, the simple programs described above could be modified as in Figure 6.1(c). After printing the message the program tells the computer to read a number from the keyboard. If this number is zero the program halts but if not, it goes on forever. It follows that how a program will proceed may depend on what data has been entered and how this affects its subsequent behaviour.

Given a particular computer program with defined input data, how might we tell whether or not it is going to halt? A possible test might be to run the program and see. Certainly, if the computer halts after some time, the question is answered but if it does not, we cannot be sure that it will go on for ever. For example, a program could be written that continually repeated some task until the date on the computer clock reached the year 3000, at which point it would halt. Examining the program's code could reveal this – and this ability to examine the workings of a computer program and tell what it will do is, arguably, a uniquely human attribute. But might there not be programs that are too complex for even a human mind to analyse successfully? Alternatively, might it be possible to write another computer program designed to examine the code and discover whether the program will halt or not? Although the latter scenario is sometimes achievable, we will see that it is impossible to devise a computer program that could examine the code of any arbitrary program and decide whether or not it will halt.

Understanding why this last statement is true involves a little more detailed discussion of how computer programs work. Let's define the expression PROG(D) to represent a computer program called PROG that is designed to read a set of data D and perform a set of operations. In the example illustrated in Figure 6.1(c), PROG represents the complete set of instructions performed. In this case, D is the single number (X) entered from the keyboard. All programming code has the form of a list, or string, of data (such as the text defining the programs illustrated in Figure 6.1) and we define P as the data string representing the code comprising PROG. Suppose it were possible to use a computer to decide whether a computer program with a given set of data is going to halt. That is, a computer program called, say, HALT(P,D), is written to accept both P and D as input data and then analyse this to determine whether PROG(D) will definitely halt, definitely never halt, or if this question cannot be answered.

The program illustrated in Figure 6.1(d) is a particularly simple version of HALT. This inputs P and D and tests whether the first line of PROG is 'STOP'. This program is so simple that it is effectively useless and could not even assess the halting properties of programs (a) or (b) but it illustrates the principle. Now consider a slightly more sophisticated version of HALT, which, after inputting P and D, proceeds to examine each of PROG's instructions in turn, looking for the statement 'STOP' or instructions that initiate an endless loop. This would successfully determine that 6.1(a) halts and that 6.1(b) does not. When applied to a program like 6.1(c), however, this version of HALT would be unable to decide one way or the other, because a branching instruction appears in the code before a STOP command or an endless loop is reached. To deal with such cases, a more sophisticated version of HALT could be devised. It would have to take into account the effect of the data (D) on PROG's behaviour and trace the course of the program through each branching instruction, looking for a point where it will definitely halt or definitely go into an endless loop.

One special but important, case is where the data inputted into PROG is its own code, P. To deal with this, we consider another hypothetical program, called TEST(P), which inputs the data string P and runs the program HALT(P,P), to test whether the program PROG(P) halts. If HALT(P,P) finds that PROG(P) halts, TEST(P) prints the message 'HALTS' and begins an endless loop. If it finds that PROG(P) does not halt, TEST(P) prints 'DOES NOT HALT' and stops. Finally, if HALT(P,P) cannot decide whether PROG(P) halts or not, TEST(P) prints 'I DON'T KNOW' and stops. Thus if TEST(P) finds that HALT(P,P) has shown that PROG(P) must halt, then TEST(P) does not halt and vice versa.

The next and crucial step is to consider what happens if TEST(P) operates on its own code (T). This case is set out in diagrammatic form in Figure 6.2, which shows that when TEST(T) runs HALT(T,T) it tries to find out whether or not TEST(T) halts. If TEST(T) finds that it does indeed halt, TEST(T) goes into an

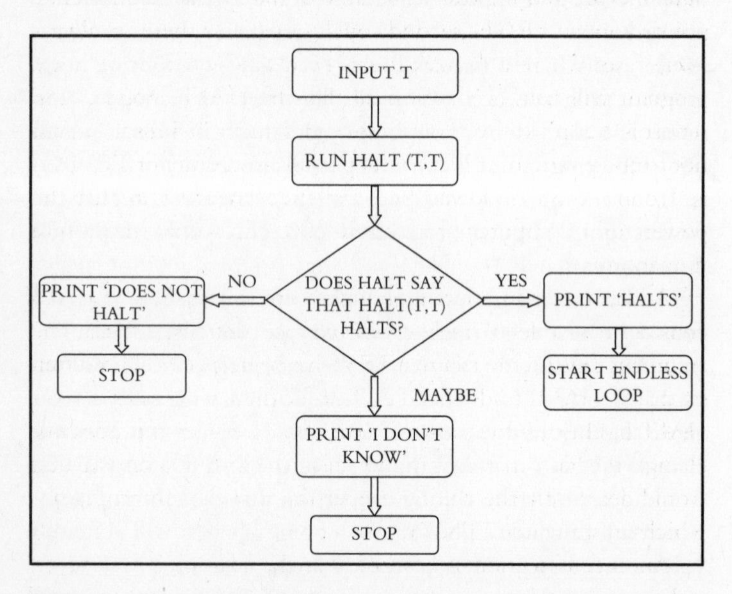

Figure 6.2 The program TEST(T). When this program analyses its own code, the only possible result is 'I DON'T KNOW' followed by STOP. (Alastair Rae)

endless loop, which is of course a logical contradiction and cannot occur if the pogram is written to follow the rules of logic. Similarly if TEST(T) were to decide that it didn't halt, it would then halt which agan is a logical contradictions. It follows that neither of these outcomes is possible, so the only possible result is for TEST(T) to say that it 'does not know' whether or not TEST(T) halts and then halt!

There are two important conclusions to be drawn. The first is that it is impossible to write a program that is capable of determining the halting properties of any arbitrary computer program – that is, one which is infallible and *never* uncertain of the result. If TEST had this property, the third option ('I don't know') would not be available to TEST(T), while the other two outcomes are also impossible, because of the logical inconsistency pointed out above. The second conclusion is that there are always some cases where a human is able to decide whether or not a program will halt, even when all that the HALT program can report is 'I don't know'. For instance, it is quite obvious to us, but not to the program, that the only possible outcome for TEST(T) is 'I don't know' followed by STOP. Does this mean that the powers of a computer program are always less than those of a human mind?

A commonly attempted rebuttal to this line of reasoning is to consider it as a deliberately contrived case – almost a trick. The argument rests on the fact that TEST has been specifically written so that if HALT finds that TEST halts, then it initiates a loop. HALT has been subject to an unfair test! However, this does not change the fact that it *is* impossible to write a program that would determine the halting properties of *any* given program – which must include TEST(T).

Another objection is to note that the reasoning we use to realize that TEST must halt when it tests itself consists of a set of logical steps. Why could these same steps not be added to HALT so that it would be able to reach a decision? This could certainly be done, but if TEST were run with this new version of HALT, the same outcome would result. Thus, even if HALT were the most powerful, robust program possible, it would still come back with the answer: 'I don't know'.

I hope you will agree that understanding the arguments leading up to this conclusion is not an especially difficult challenge for a human mind. This may imply that this type of understanding is different in nature from any that may be possessed by a computer program. But that is not necessarily true.

First, suppose that the brain operates computationally – that all our thinking could be simulated by the operation of a computer program. It follows that one of the things our brains can do is to examine other computer programs and decide on their halting properties; when doing so it would essentially be operating a version of HALT – call this HALTME. If the details of HALTME were known, they could be inserted in TEST(T), just as before, which would therefore halt while saying that it did not know whether or not it would halt. Thus, absurdly, we would know that TEST(T) was going to halt but HALTME would not. So, at least one of the assumptions we've made above must be false. Roger Penrose concludes that this falsifies the initial premise that the brain acts in the same way as a computer.

Penrose's ideas have been strongly criticized, particularly by AI researchers. One line of criticism focuses on the fact that TEST(T) may not be a 'sound' program in all cases. If the HALT program could itself be mistaken when it decides that a PROG(D) halts, then it may also be mistaken when it examines its own code. The question is 'How can a computer engineer determine whether a given version of HALT is sound?' A person might examine the code but how does she know that she has done so reliably? Imagine that this argument is applied to my mind. How do I know that HALTME is sound? One thing I know about myself is that I am perfectly capable of making mistakes. The human mind is very complex – much more complex than any computer program yet devised. If I were able to analyse my mind to the point where I could write down its reasoning processes in the form of a succession of operations, I wouldn't be surprised to find that its conclusions were not always reliable. Penrose's answer to this is to say that it is unlikely that a faulty program would be more powerful than a sound one but he gives no argument to support this.

The halting properties of computer programs are just a particular example of a more general result obtained by Kurt Gödel, the Austrian logician, mathematician and philosopher. In 1931, he shocked the mathematical world when he showed that it was impossible to derive all the truths of arithmetic from any given set of consistent axioms and the rules of logic. However large this set of axioms, there is always at least one true statement that cannot be derived using them alone. The proof of Gödel's theorem is more complex than that for the halting problem but it also relies on applying the theorem to itself. Indeed, Gödel's theorem can be shown to follow from the fact that no computer program can overcome the halting problem. Once again, it seems that the human mind may be able to understand an argument that could never be understood by a machine whose operation relies on a finite set of rules.

Penrose's conclusions, though based on quite different reasoning, are similar to those reached by Searle in the Chinese room example. Both conclude that 'understanding' reaches beyond the abilities of a machine or brain that is based only on computational principles. Understanding is a vital part of human thinking in nearly everything we do and appears to be a uniquely human attribute. It would be hard to see how one might explain the halting problem to a dog or a chimpanzee.

Penrose goes further, however. He points out that, at a fundamental level, the laws of physics predict that the physical world should proceed in a logical manner – the technical phrase is that the laws are *computational*. This implies that all physical events are fundamentally equivalent to the steps in some computer program, which, although unimaginably enormous and complex, could still not overcome the halting problem. He believes that this implies that some form of new physics is needed to explain consciousness. One suggestion he has made is that the brain has evolved to contain within it structures that are capable of operating as a *quantum computer*, a theoretical form of computer that employs features of quantum physics to perform some particular calculations much more powerfully and rapidly than any conventional computer. Today, there is no convincing evidence that the brain is capable of acting in this way or that a quantum computer could overcome the contradictions that lead to the halting problem. However, even though Penrose maintains that consciousness cannot be simulated using a conventional computer, he still believes that the phenomenon supervenes on the physical brain. If he is right in this belief, the principles of reductionism should continue to apply.

The three worlds of Karl Popper

We first met the philosopher Karl Popper in Chapter 1, where I discussed his work on the principle of falsifiability. Popper made important contributions to various areas of philosophy and, towards the end of his life, turned his attention to the problem of consciousness. Unlike John Searle and Roger Penrose, who both reject Cartesian dualism, despite their reservations about the computer as a model of the brain, Popper believed that there is a 'ghost in the machine'. He shared this view with the Nobel Prize-winning brain scientist Sir John Eccles, with whom, in 1977, he wrote the book *The Self and its Brain*. In the first section of this volume, Popper produced philosophical arguments that led him to define three types of object which he called 'world-1', 'world-2' and 'world-3'. World-1 objects are those that comprise the whole of physical reality – solids, liquids, and gasses, along with their component atoms and fields. In contrast, world-3 objects are not themselves physical but are things like stories, pieces of music, mathematical theorems, scientific theories, and so on. World-2 consists of the states of consciousness in the brain that provide an interface between world-1 and world-3.

Popper argued that world-3 objects must be considered to have reality, because their existence leads to changes in the physical universe, that is, in world-1. Consider a piece of music. What is it? It is certainly not the paper and ink used to make a copy of the score, neither is it the CD on which a particular performance is recorded; it isn't even the set of sound vibrations in the air when the music is played. None of these *are* the piece of music but all these world-1 objects exist in the form they do *because of* the music. Another world-3 object is a mathematical theorem, such as 'The only even prime number is 2'. Everyone who knows any mathematics must agree that this statement is true. It follows that it is 'real' if only because world-1 objects, such as the arrangement of the ink on the paper of this page, would otherwise have been different.

The next stage in Popper's argument is to note that the reality of world-3 objects depends on the existence and action of a conscious human mind. The piece of music or the mathematical theorem results in a particular mental state of a human being (a world-2 object), which in turn affects the behaviour of world-1. Without human consciousness this interaction would be impossible and the reality of world-3 could not be established. This led him to conclude that the self-conscious human mind must also be real: he argued that, as only a self-conscious human being can appreciate the reality of world-3 objects, human consciousness itself must have reality beyond that of any physical object, *even the brain*.

Popper's ideas were developed further by John Eccles in a later section of the book. Having described the physiological operation of the brain, he speculated on how the self-conscious mind interacts with the world-1 brain. His approach is reminiscent of that of Descartes but instead of the pineal gland, Eccles postulated that there are special 'open synapses' in the brain that are directly affected by the non-physical, self-conscious mind. Both Popper and Eccles explicitly stated that they believed there is 'a ghost in the machine', meaning an aspect of mind that cannot be encompassed solely by the physical brain – in other words, a 'soul'.

Is it true that products of the human mind, such as pieces of art and music, in some sense exist beyond the physical world? Why should we assume that a piece of music has an existence beyond its score and the instances of its performance – live or recorded? When we hear a performance, we may be moved by its great beauty; even say that it 'stirs our soul' but why is this not another example of the ability of our brains to react to the sensory inputs it receives? If we hear another version of the same piece, we might think that it is of a higher or lower standard than the previous performance and find, drawing upon our memory, that it gives us more or less pleasure. But this is not to say that the piece of music itself has an existence beyond all the instances of its performance.

Is mathematics real?

The example of the mathematical theorem presents somewhat more of a challenge. The fundamental nature of mathematics has been hotly debated by philosophers for a very long time. In its simplest terms, the controversy focuses on whether mathematical truths are discovered or invented. Are mathematical theorems 'out there' – in some sense part of the universe – or are they purely products of the human mind? Originally, at least, mathematics dealt with numbers, so let us consider whether *number*, as a concept, exists in the universe or is a cultural product like art or music. When they first encounter this question, many people seem certain that numbers exist in and of themselves, since they can be used to count discrete objects, ranging from apples to galaxies. Surely the very concept of number and its properties must transcend the particular examples it is applied to. Properties of numbers, such as the fact that they can be added or multiplied together, or the uniqueness of the number two as an even prime, are surely things for us to discover rather than invent.

There is another point of view, however. Perhaps we invented the concept of *number* for utilitarian reasons: we needed some abstract objects with properties that would help us organize our description of certain features of the physical universe. We quickly found that the 'natural numbers' (that is, the positive integers 1, 2, 3, and so on) could not do everything we wanted to do: for example, to measure sizes and weights, fractions are needed. 'Real numbers' were invented, which assume that there are infinitely many (another concept!) numbers between every pair of successive whole numbers. When we measure something physical, we can only express the answer to so many decimal places; we may imagine that all the other infinite possibilities exist but if we do, they are our mental constructs. The concept of number has also been extended beyond the natural and real numbers to include negative numbers and what are known as *imaginary numbers*, because we have found these to be useful in developing our understanding of the world around us.

Negative numbers, for example, were not thought of until a long time after the development of arithmetic. Previously, some mathematical descriptions were written out twice, once for distances measured to the right and then again for distances measured to the left. Much later, it was discovered that these could be unified into a single description, if right- and left-pointing numbers were respectively termed positive and negative. Moreover, this works only if negative numbers obeyed certain rules, such as 'minus times minus equals plus', which are routinely taught in schools but are not necessarily intuitively obvious. Many believe that this is evidence for the idea that mathematics is an invention of the human mind. If negative numbers had been waiting 'out there' for some unconscious mind to make contact and discover them, as Popper and Eccles argued, mightn't this have happened sooner and more directly?

Not everyone is convinced by this argument – and the nay-sayers include Roger Penrose. What might decide the issue? One answer would be if a computer following what were known to be logical steps acquired an intelligence that was able to come up with a mathematical theorem – for example invent negative numbers or show awareness of the uniqueness of the prime number 2 – without having been given this information in advance. Presumably the computer would not have 'open synapses', so it would not be able to make contact with the world of mathematics 'out there' and could only have obtained these results by a process of invention, rather than discovery.

Such a computer might also be able to pass the Turing test and understand the halting problem, despite Penrose's argument that this is impossible. There is a (probably apocryphal) story that some scientists once proved that an insect like a bumblebee should be unable to fly. The fact that it can actually do so might be thought to falsify the laws of aerodynamics but the truth is that those who denied the possibility of bee flight had shown only that the bumblebee cannot glide through the air; they had not taken into account the fact that a bumble bee continually flaps its wings in order to stay aloft. If and when a computer is developed that can prove mathematical theorems and understand the halting problem, I suspect we shall discover that it is operating in a way that we just hadn't thought of.

You might well ask why I have given so much attention to mathematics in a chapter on the mind. After all, maths plays a very small part in most people's lives, let alone their thinking. Indeed, I have used very little mathematics in this book because

I am hoping that I have been able to provide explanations that are accessible and interesting to as many people as possible. Despite this, I suspect that a significant number of readers have found it quite challenging to understand the arguments relating to the computer halting problem. Does this mean that people with mathematical ability have a special skill that allows them to understand an argument that is beyond the capacity of a machine but that everyone else's ability is limited by the rules governing the operation of computers? I think that most of us – mathematicians and non-mathematicians alike – would need a lot of convincing before we were ready to accept that hypothesis.

The conscious mind is able to undergo experiences and perform tasks that are astonishingly complex and amazing. Despite the challenges discussed in this chapter, the possibility that the mind, with all its properties, supervenes on the operation of the brain and its complex network of neural connections has not been falsified. The properties that emerge from the brain are more than those possessed by its component parts – just as water and ice have properties beyond those of hydrogen and oxygen. Francis Crick, one of the discoverers of DNA, once said: 'human beings are essentially a bag of neurons'. But the neurons in our brain are not just 'in a bag'; they are connected in a hugely complex manner. It is the pattern of these connections and their development over time that allows our experiences, conscious and unconscious, to supervene upon them. In this sense, we are indeed more than Shakespeare's 'quintessence of dust'.

7 Society and the individual

Sometimes, when we look across a town or city in the early evening, it seems that a plume of smoke is emerging from some of the buildings. It appears, builds quite rapidly and dies away again over a period of half an hour or so, just as if a fire had started and been put out by a vigilant fire service. If you were in the town at the time, however, you would have seen not flames but a host of starlings coming to roost. When the flock arrived, it appeared to form a cloud above the buildings and as individual birds found places to land, the mass steadily reduced in size until all the birds had settled for the night. If during this process, you focused your attention on one particular starling, it would be seen to be following the motion of the flock but at the same time flying this way and that – perhaps to avoid collision with other birds, perhaps to seek a good roosting spot.

This is an example of flocking behaviour, which is exhibited by many species of birds, not just when coming in to roost but much of the time. Typically, large numbers of birds move together across the sky. Individual birds appear to fly in the same direction at the same speed and the whole flock can change direction, apparently in unison, wheeling and banking almost as if it were a single creature. Biologists believe this to be an evolutionary strategy that helps protect members of the species from predators. By roosting, feeding, and migrating in flocks, only a minority need be on the lookout for danger, while the others concentrate on the business in hand. In the event of an attack, the watchers give the alarm and the flock quickly moves to minimize danger. This does not provide one hundred percent protection for every bird, of course, but a lurking hawk is not likely to be able to kill many members of the flock. Furthermore, there is safety in numbers. If the birds flew in random directions, rather than together, they would be an easier target. Not only birds behave in this way. Many species of fish move in shoals and herds of animals subject to predators – think of antelopes hunted by lions – may stampede in a similar fashion.

How does such co-operative behaviour arise? First, suppose we wish to get a number of people to move collectively as a group. One way to do this would be to appoint a leader who could continually pass instructions to each individual, telling them what to do – by shouting, or perhaps by using some form of radio communication or signalling flags. This might work well if the instructions were simple and much the same for everyone, as in the case of a platoon of soldiers exercising on a parade ground. The leader orders 'right turn', 'halt', and other commands; if everything goes according to training, the whole troop responds in unison. But therein lies the trick: to be effective, such behaviour has to be drilled into the group by repeatedly practising the same manoeuvre over and over again (that's why the leader in basic training camps is called the 'drill sergeant'). Only after the responses become routine can a spectacle such as Trooping the Colour in front of the monarch be performed with the degree of perfection that we have come to expect.

It is very hard, however, to see how flocking behaviour could be explained through communication and training of this sort. There is no evidence that a flock is controlled by a 'leader' bird and no known mechanism by which any such bird could convey moment-by-moment instructions to its 'troops'. Moreover, achieving the variety of flocking manoeuvres exhibited by birds soaring, banking, and diving in unison would require a huge amount of practice drill. Nevertheless, before the 1980s, when flocking behaviour was largely a matter for speculation and controversy, it was seriously suggested that birds in a flock must be able to communicate with each other using some unknown additional sense – even perhaps supernaturally, using extrasensory perception.

Another conceivable explanation might be that each bird instinctively knows exactly what it will be doing. This would mean that somehow all the birds in a flock have detailed instructions programmed into them from birth – presumably encoded in their DNA from generation to generation. When a bird was confronted with a certain type of sensory input – say, a predator approaching from below, it would recognize that it needed to soar and its brain would activate a 'soaring program'; when it needed to bank, the 'banking program' would be set in motion. Still, this would not explain why birds do what they do *in unison*, especially in response to an input such as a lurking hawk. After all, the hawk would not be in the same position in relation to all the starlings in a flock – so some would follow the instructions to soar while others were supposed to bank. We would then see individual birds flying off in different directions – not flocking. If somehow this were the case, it would mean that flocking behaviour was confined to only a few possible manoeuvres, where acting in unison would be the appropriate response for each and every bird in the flock. Yet all evidence indicates that the motion of the flock as a whole is very varied, allowing it to react to many different situations. Its only predictability is that whatever the flock is doing, all the birds do it together.

The actual mechanism for bird flocking is now believed to be quite simple and at the same time perhaps more surprising than any of these hypotheses. It appears that flocking behaviour emerges naturally and inevitably because each bird moves in response to the behaviour of the other birds in its immediate environment, following a set of simple rules. In 1986, while working for the entertainment company Symbolics Graphics Division, Craig Reynolds (later a senior researcher with Sony) developed a computer simulation of flocking behaviour. The program calculated what the motion of a set of birds would be if the actions of each were determined by three principles: *separation, alignment,* and *cohesion*. Each bird was assumed to be aware only of its immediate neighbours who are flying in front, above, below, and on both sides, but not behind it. To achieve separation, each was programmed to steer so as to avoid colliding with any of its neighbours; to achieve alignment, each bird tried to fly in the same direction as its neighbours; and to achieve cohesion, each bird aimed to move towards the centre of its neighbourhood group. The computer calculated the motion required for a particular bird to achieve each of the three aims. The movement of the bird was then taken to be the average of these figures; the computer program then repeated the calculations for every bird in the simulation. Thus, it is assumed that every bird responds only to its local environment; no single bird has knowledge of what more distant birds are doing or awareness of the motion of the flock as a whole. There is no leader bird.

When Reynolds ran his computer simulation, he found that he was able to reproduce all the major features of flocking behaviour. His computer birds moved like a flock of real birds. Indeed, in the scientific paper reporting his work, he says that 'many people who view these animated flocks immediately recognize them as a representation of a natural flock and find them similarly delightful to watch . . .' The model was further tested by inserting obstacles in the path of the flock: the flock divided as it approached each object then reunited after it had passed, closely mimicking the behaviour of a flock of real birds encountering a physical barrier such as a tower or a chimney. At the time of writing, this simulation could be seen at http://www.red3d.com/cwr/boids/.

In the twenty or so years since Reynolds's breakthrough, many other successful simulations of animal behaviour have been developed. These now form an important part of our understanding of the natural world. Of course, computer simulation does not *prove* that the flocking behaviour of real birds is caused in the same way as that of the computer birds – it is more the equivalent of an avian Turing test – but it provides very strong support for such a hypothesis. It is perfectly plausible that birds should have evolved with an instinct to follow rules similar to those programmed for the computer birds.

Although the computer program demonstrates the power of the model, we still have not answered the question of *how* it works. If a flock were moving in a straight line, it is reasonably obvious that the rules would ensure that the individual birds all follow the same path but what happens when the flock executes a manoeuvre – for example, a sharp right turn? To understand this, remember that although each bird is following the rules, there is still some randomness to its motion: while the flock is moving in a particular direction, there is a chance that a particular bird might start flying off course. In most cases, it will see what its neighbours are doing and correct its behaviour, bringing it back 'into line'. However, on some occasions a neighbour will start following the deviating bird; its neighbours may then copy this motion and so on until the whole flock begins to fly in a new direction.

Flocking behaviour is a clear example of the application of the reductionist approach. The flock is a higher-level object, which displays properties that are not possessed by the individual, lower-level objects – the birds. We can *reduce* a flock to a set of individual birds, which obey only lower-level rules. The behaviour of the flock as a whole then emerges as a result of bringing the birds together: the rules that describe the higher-level behaviour supervene naturally on those operating at a lower level.

Human behaviour

The next question is whether the collective behaviour of human beings also follows rules that emerge from the behaviour of the individuals within a group. Can reductionist principles, similar to those operating in the case of birds, be applied to the behaviour of groups of humans – including groups as large as the society that comprises, say, a marketplace or a modern nation state?

Anyone who has driven (or indeed been a passenger in) a motor vehicle travelling on a motorway or other arterial road will almost certainly have experienced the phenomenon that has come to be known as a 'phantom traffic jam'. Imagine that you are travelling in a vehicle on such a road. The traffic is heavy but moving at a reasonable pace when, often quite suddenly, the vehicle in front slows down, causing yours and other cars behind to do the same. After a short time, you, and a number of vehicles in front and behind, are forced to come to a halt. After remaining stationary for several seconds or even a minute or two, the car in front starts to move and you resume your journey, soon recovering your original speed. You wonder why this has happened but there is no sign of road works, an accident, a police check, or indeed anything else that could have been expected to cause this hold-up. After driving further, however, you are again brought to a halt for a short time before once more moving on; this pattern may repeat itself a number of times.

Someone observing this from a height – say in a helicopter hovering above the road – will see a long line of vehicles, some of which will be moving quite freely but others, in one or more sections, will be at a standstill. The stationary vehicles are much closer together than those that are moving. As time goes on, this stationary section is seen to move back through the line of traffic, affecting different sets of vehicles. This movement has the appearance of a wave moving along the line of traffic, with the wave peaks corresponding to the stationary regions, where the vehicle concentration is at its greatest, and the troughs to the most-freely moving part, where the vehicles are further apart.

The point I want to stress about this phenomenon is that none of the drivers involved has deliberately planned for it to happen. Each motorist is attempting to drive as quickly as possible, consistent with the need to avoid collisions with other vehicles. The wave-like behaviour emerges spontaneously, as has been confirmed experimentally in deliberately created scenarios and by observations of actual traffic on public roads. It also emerges from computer simulations and mathematical analyses of traffic flow; you can easily find examples on the Internet by entering 'phantom traffic jams' into a search engine. Provided the traffic density is sufficiently high and is attempting to move fast enough, the randomness resulting from small differences between the behaviour of different drivers – the time they take to stop, the speed they aim to drive at and so on – is sufficient to trigger wave motion. It has been found that two prerequisites must be fulfilled for this behaviour to emerge: high traffic density and a high maximum speed when the traffic is free-moving. These factors are connected – a lower maximum speed is needed if the density is high and vice versa.

In this context, human beings are behaving rather like starlings. Just as the birds fly along, affected only by the other birds that are close to them, drivers aim to travel along the road as quickly as they can, influenced only by the motion of the vehicle immediately in front. Both the starlings' collective flocking motion and the phantom traffic jams emerge naturally from the behaviour of the component individuals who are just minding their own business.

The study of traffic flows has shown that the average rate of traffic flow actually increases if vehicles maintain a lower but steady speed, instead of the stop-start motion associated with phantom jams. Two strategies have been implemented to try to encourage this. One is the reduction of traffic density, by building additional roads or adding more lanes to existing roads; another is the introduction of lower speed limits (typically 50 mph when the free-moving maximum is 70 mph) at busy times, which actually increases the average traffic speed. This is where human behaviour is very different from that of starlings. The ability of some of its members to observe and understand the collective behaviour of human society has developed to the point where regulations, such as speed limits, have been introduced. This type of intervention is what makes the behaviour of human society so much more complex than that of birds or animals. The individual drivers in a phantom jam (most of them anyway) are 'doing their own thing' in much the same way as are the starlings in a flock or indeed, the electrons in an atom but this behaviour can be modified by the imposition of human inventions, such as speed limits.

This crucial difference between human beings and birds is why the scientific study of human society has been found to be so challenging. At least some of its members have some understanding of how their society behaves as a whole and how it can be affected by their actions. Given the power to do so, this understanding can be used to generate laws and regulations that act at the 'higher level' of society as a whole. Consider, for instance, when a parliament passes a law that defines some set of actions as criminal. Calling some behaviour a 'crime' is a way to help ensure that individual people are discouraged from doing it. The crime is announced to society and a police force is given instructions to watch out for actions that would be categorized as criminal. When people are discovered to have behaved in this manner, they are normally punished in some way. Those making the laws, whether a democratically elected parliament or an absolute monarch, do so believing that they understand what their consequences will be for society as a whole as well as its individual members. Laws can still be seen as 'emerging' from the behaviour of its individual members but they have been deliberately and consciously planned by the particular individuals who have the power to make them.

All our previous applications of reductionism have featured higher-level rules emerging from a lower-level substructure whose members have no awareness of the higher-level outcomes. The properties of a molecule are not contained in any of its component atoms but emerge as they interact. Ice forms when many water molecules are brought together at a low enough temperature but its properties, such as solidity, are not possessed by the individual molecules. Similarly, individual neurons are subject to basic biochemical laws and have no awareness of the fact that they will create a brain capable of instinctive motor actions, sensory processing, thought, and even reasoning and consciousness. Individual starlings have no knowledge of the fact that the rules they are following will lead to the behaviour displayed by the flock as a whole. Human beings, in contrast, do have some understanding of how they behave collectively.

Yet, once these laws are in place, it could be said that each member of society acts rather like a starling in the sense that an individual obeys the rules (or, indeed breaks them) for their own reasons, rather than because of the effect their actions may have on society as a whole. The designers of, say, the law against theft aimed to produce a society in which both privately and publicly owned property can be safely protected, As a result, many of the individuals who choose not to steal do so because they judge that the likelihood of detection and punishment is greater than that of successfully accomplishing the theft. They may be concerned with whether their choice to refrain from stealing is good for people living on the other side of town or the other side of the country but this is certainly not always the case. Indeed, if such motivations were universal, there would be no need for any laws!

Another example of an emergent phenomenon is the patterns of behaviour or beliefs that are common to all or some of the members of a society. These behaviours can spread through the society from one group to another: for example, a pop song may be broadcast in one country and a few days later, sung and enjoyed worldwide. The biologist Richard Dawkins termed such phenomena *memes*. He saw the process of meme transmission as closely analogous to the reproduction of biological organisms through their genes. The meme – or the gene – can be altered in response to its environment, although in the case of memes this is often by the inheritance of acquired characteristics, which, as we saw in Chapter 4, does not occur in Darwinian evolution. Memes are often compared with viruses, because they can spread rapidly through a population, 'infecting' its members. This language is now commonly used, such as when we talk about a message or video 'going viral' on the Internet. The concept of the meme is quite controversial, particularly when it is applied to political and religious beliefs: some people find the comparison of the spread of a religion with that of a virus offensive.

The social sciences

The social sciences are concerned with the study of society. They encompass subjects such as anthropology, criminology, history, geography, and politics, which all aim to ascertain and explain traits of collective human behaviour. The social science of economics is dedicated to describing the rules governing the production, distribution, and consumption of goods and services by humans. The founding father of economics is generally accepted to be Adam Smith, who was born in 1723 and died in 1790. He was one of the key members of the Scottish Enlightenment, which also included the philosopher David Hume and the poet Robert Burns. Smith is today best known as the author of *An Inquiry into the Nature and Causes of the Wealth of Nations,* published in 1776, which is generally considered to be the first modern work of economics and still influences many thinkers.

The Wealth of Nations contains a number of original insights into the nature of the industrial society that was developing rapidly in Britain at that time. One of these is the idea of *division of labour*. In pre-industrial society, goods were produced in small domestic workshops, the whole process being carried out by a small number of people, who were often members of the same family. To produce more complex manufactured goods, production moved into factories, where workers carried out different specialized tasks. This eventually led to the invention of the production line, which typified manufacturing in the twentieth century. Under division of labour, workers only had to carry out their allotted task. They did not need to have any knowledge or awareness of the nature of the finished product or of the full manufacturing process. It was also assumed that their main motivation in being a part of the process was to further their own material interests – by working for wages or, for those of a higher station in society, investing their capital in an enterprise.

In a widely quoted passage from *The Wealth of Nations*, Smith generalized these ideas to describe the operation of the economy as a whole:

> As every individual, therefore, endeavours as much as he can both to employ his capital in the support of domestic industry and so to direct that industry that its produce may be of the greatest value; every individual necessarily labours to render the annual revenue of the society as great as he can. He generally, indeed, neither intends to promote the public interest, nor knows how much he is promoting it. By preferring the support of domestic to that of foreign industry, he intends only his own security; and by directing that industry in such a manner as its produce may be of the greatest value, he intends only his own gain and he is in this, as in many other cases, led by an invisible hand to promote an end which was no part of his intention. Nor is it always the worse for the society that it was no part of it. By pursuing his own interest he frequently promotes that of the society more effectually than when he really intends to promote it. I have never known much good done by those who affected to trade for the public good.

Nowadays, Smith's concept is often given the name 'free-market economics', because it states that if every individual in society is free to act so as to further their own material interests, this will result in economic growth for the society as a whole, with an increase in the wealth of the nation as the result. Accepted at face value, the laws regulating the economy emerge like an 'invisible hand' from the behaviour of a society's individual members. People follow the laws that apply to them without any awareness of their social effects. In this sense, human beings are behaving like starlings pursuing their own perceived interests without any awareness of the effect this has on society as a whole.

Free-market economics has considerable backers among some academics and politicians – indeed, it is widely believed that Margaret Thatcher, the former UK prime minister, carried a copy of *The Wealth of Nations* in her handbag. The ideas Adam Smith sets out in the quote above are, however, deeply controversial. Nearly everyone – including Smith himself, judging from later passages in the book – believes that this freedom must be complemented by at least some regulations. That is because human beings, unlike starlings, can build an understanding of what is happening at the level of society as a whole. They have to be prevented from using this knowledge to exploit the free market to their own ends.

The nineteenth-century German philosopher, Karl Marx, is often held up as the diametrical opposite in thinking to Adam Smith. Marx concurred, however, with Smith's idea of a division of labour and that free-market capitalism would lead to growth in an economy's productive capacity. Where Marx's conclusions differed from Smith's was in how these principles would affect certain groups of individuals in society. In Marx's view, free-market capitalism would inevitably lead to the increasing exploitation and impoverishment of those whose labour was necessary to the basic productive process – the 'working class' or, as he called them, the *proletariat*.

Marx and his fellow German political theorist Friedrich Engels, with whom he wrote *Das Kapital*, were strongly influenced by the poverty and deprivation suffered by the industrial workers of their time. They predicted that wealth would become concentrated in fewer and fewer hands, that workers would become increasingly impoverished and in turn fewer people would be able to buy the products of industry. This cycle of events would lead to the collapse of the free-market capitalist system and they proposed that it would eventually be replaced by a 'communist' society. One stage along this road from capitalism to communism would involve the establishment of a 'socialist society' in which all large-scale enterprises would be carried out in 'socialized units' – although it's not entirely clear what they meant by this.

After the Russian revolution, there was an attempt to create a socialist society – the Soviet Union. The political leaders chose to implement Marx's idea by having all the means of production, distribution, and exchange owned and operated by the state, which was governed by a 'dictatorship of the proletariat'. Here, the laws governing the economy would not emerge from the behaviour of individual members but be devised and enforced by the organs of the state to ensure that everyone received a fair reward for their labour. These laws would be nothing like the flocking behaviour of starlings; they were more like an officer overseeing the Trooping of the Colour. The higher-level laws were planned and imposed from above with the conscious aim of advancing the 'common good'.

The socialist system (at least as interpreted and implemented in Eastern Europe after World War II) implies a command economy, where everything is managed by the state, but this is not a necessary part of Marx's prescription. In any case, Marx saw any such arrangements as only temporary. He believed that the overthrow of capitalism would lead to a change in people's consciousness, such that the laws governing their behaviour would also change.

People would naturally become more concerned with society as a whole. They would refrain from exploiting laws for their own benefit, for example by stealing, because that would be good for them as well as people living on the other side of town. State control would no longer be necessary. The state would 'wither away' and, just as in a free-market economy, individuals would follow their personal inclinations. These would, however, be different from those pertaining under capitalism; a system of production and consumption would emerge to ensure the implementation of the guiding principle: 'from each according to his ability, to each according to his need'. According to Marx, such a society would be rich and its wealth would be used to further the quality of life of the individual:

> In communist society, where nobody has one exclusive sphere of activity but each can become accomplished in any branch he wishes, society regulates the general production and thus makes it possible for me to do one thing today and another tomorrow, to hunt in the morning, fish in the afternoon, rear cattle in the evening, criticize after dinner, just as I have a mind, without ever becoming hunter, fisherman, herdsman or critic.

Thus, Marx's vision of a communist society could be described as one in which individuals follow their own interests but these interests are different from those they would perceive in a capitalist context.

In such a society, people would behave like starlings but the rules that they naturally follow would have changed and higher-level laws would emerge naturally to ensure that society was organized to further the common good. Whether human nature can be altered as a response to a change in the environment is a deeply controversial idea but it is not only held by committed Marxists. In the 1990s, some of the supporters of implementing a free-market economy in Eastern Europe believed that their difficulties in making the transition were due to the years of training that individuals had received. They had been taught to think in collective terms rather than as rational individuals seeking their own interests!

The ideas of Smith and Marx represent the extremes of free-market capitalism and communism. Since the nineteenth century, a number of alternative economic theories and practices have been suggested. Notably, the British economist John Maynard Keynes proposed that if the state borrowed and spent money at times of recession and withdrew money from the economy during boom times, this would allow free markets to operate to everyone's advantage. Keynes's ideas were especially influential during the Great Depression, which ravaged the global economy in the 1930s and influenced President Roosevelt's New Deal.

Despite the apparent success of Keynesian policies then and during World War II, the extent to which the state should be involved in managing the economy is still a matter of considerable debate. Those on the 'right' argue for the benefits of a free market with minimum regulation, while 'left-wing' economists see the virtues of state intervention aimed at ensuring efficiency and fairness. In many situations, the differences have become blurred, with many on the right recognizing an important role for the state and many on the left seeing virtues in individual choice. For example, communist China has in recent years mostly confined the government's role to the control of the major industrial and financial instruments of the economy – banks in particular – while leaving the rest of the economy to run on the free-market model. It has been controversially argued that this means that the current Chinese system is closer to the socialist society envisaged by Marx than it is to the Stalinist model that was followed in Eastern Europe.

Adam Smith's version of the free-market economy and Marx's vision of the ideal communist society both assume that their desired features of society as a whole will emerge from the natural behaviour of individuals. Smith assumed that everyone would act rationally to further their material wealth, while Marx believed that the transformation to a communist society would inspire individuals to act naturally for the common good. The latter principle will remain untested unless and until such an ideal communist society emerges but the former claims to be a description of how people generally behave today and can therefore be tested directly.

One claimed benefit of the free market is stability, particularly of price. If a seller offers to sell something at a lower than average price, buyers will seek to take advantage of this, a shortage will develop and the seller will respond by raising the price. Similarly, a higher than average price will result in a surplus and a price reduction. It is claimed that if everyone involved seeks their own economic advantage, this will result in the prices of all the goods in the market settling down to stable values. Recent research using computer simulations and careful analysis of the behaviour of real markets has, however, shown that, although stability emerges in many situations, this is not universal. Even when everyone is following the rules, instabilities can arise where prices become unstable, increasing or decreasing by huge amounts. A recent example was the worldwide banking crisis triggered by the housing market in 2008. This property of markets is somewhat analogous to that of a flock of starlings flying in one direction and then changing course suddenly for no apparent reason.

In addition, there is considerable evidence indicating that Smith's assumption that individuals will act so as to optimize their material wealth is only a partial explanation for the way most individuals behave. An illustration of this can be found in the 'Ultimatum' game that was invented in 1982 as part of research into how individuals make economic choices. This very simple game involves two players: the 'proposer' and the 'responder'. A 'controller', usually one of the researchers carrying out the experiment, monitors the game. To begin, the controller gives the proposer ten coins, which represent a unit of currency, with the instruction that he is to offer some number of coins to the other player. If the responder accepts the proposer's offer, both players are allowed to keep the money they now have, but if the responder refuses, all the money is returned to the controller. Under the economic self-interest suggested by Adam Smith, both players should act rationally and they should also expect the other player to act rationally. Looked at from such a perspective, the best strategy is for the proposer to offer the responder as little as possible – that is, one coin – to maximize the amount of money he will have at the end of the game. A rational responder will accept this offer, since she has no coins at the start of the game – one coin is better than nothing. Each player will have gained. If she refuses, both players lose everything.

When real people play this game, however, they are found to behave quite differently. If the proposer's offer is low, the responder usually rejects it, presumably considering the proposer's actions to be so unfair that she is prepared to make a sacrifice to ensure that the proposer is punished by losing everything. This game has been played by subjects around the world and in almost all cultures the minimum acceptable offer is around four coins – 40 percent of the total.

The Ultimatum, and other similar games known as *behavioural economics*, has been carried out while the players' brains are being observed using fMRI scanners. Researchers found that a low offer from the proposer triggers a response in the part of the responder's brain that is associated with negative feelings, such as pain or disgust, while the act of punishing the proposer by refusing the offer initiated firing in the brain's pleasure centres. The tendency of human beings to seek fairness seems to be deeply embedded in our psyche, apparently pre-programmed into our brains, rather as starlings are pre-programmed to fly in flocks.

The relevance of such games to the behaviour of the more powerful members of society may be limited for two reasons.

First, becoming powerful may involve ruthlessness, which could mean that such people would play the game in an atypically selfish way. Second, most people might act differently if Ultimatum were not a game but real life – with a lot of money (or something more valuable) at stake. If the original pot of money given to the proposer were one million dollars, how many responders would refuse an offer of one hundred thousand dollars in order to deprive the proposer of nine hundred thousand? Unfortunately, the likelihood of obtaining the research funding needed to test this out on a significant sample of people must be very small!

Can an understanding of human nature be used to devise an economic theory that would correspond to reality better than Adam Smith's free market? A new field of investigation, neuroeconomics, has grown up in recent years in an attempt to do just that. Some neuroeconomists hope to use our increasing knowledge and understanding of how the human brain works to devise sets of laws and regulations that will lead to desired economic ends. Of course, this begs the question of what these ends should be: is it economic growth (that is, an increase in the 'annual revenue of the society', using Smith's phrase) or should it take into account actions needed to protect the environment, for example? Might a way be found to produce something akin to Marx's ideal communist society? If such tools do emerge from the study of the human brain, the question of how this knowledge should be applied, and to what ends, will remain. Such choices are of course political: one possibility is that they be subject to some form of democratic decision-making but another is that the experts in this subject will end up advising the rulers of society on how better to control the rest of us.

Summarizing, there is no evidence that the reductionist principle fails to apply at the level of human society. In some situations, such as phantom jams, collective behaviour emerges from the actions of individuals in a reasonably straightforward way. In general, however, this is greatly complicated by the capacity of individuals to understand what is happening at the collective level and to adjust their own actions accordingly. Given that this happens, however, the idea that societal behaviour supervenes on that of the individuals within it has not been falsified.

We have now reached the point where reductionism, based on the principles of falsifiability, simplicity, and emergence has been shown to apply all the way from the internal structure of the atom to the collective properties of society. In doing so, however, we have bypassed an area where, arguably, reductionism is most seriously challenged. The next chapter aims to put this right.

8

Reducing the quantum

In the early years of the twentieth century, two radically new ideas entered the world of physics. One was Einstein's theory of relativity and the other was quantum physics. We met some of the ideas of quantum physics in Chapter 2, where we saw that a physical system, like an electron within an atom, has an associated wavefunction whose intensity at a particular point determines the probability of finding the electron in that vicinity. Because the negatively charged electron is attracted to the positively charged nucleus, it is confined to a small space around it into which the wavefunction has to fit. This leads to restrictions on the possible shapes and sizes of the allowed wavefunctions, which in turn, mean that the electron's energy can take on only particular, fixed values. All this can be analysed mathematically using the Schrödinger equation and the results are in complete agreement with experiment. Moreover, a similar analysis can be successfully used to explain how other properties of atoms and molecules, which form the material physical universe, are also explained by quantum physics.

In this chapter, I am going to discuss the fundamental ideas underlying quantum physics in more detail, concentrating on the underlying conceptual principles that challenge assumptions about the natural world which are easily taken for granted. Indeed, some implications of this, including the now famous (or notorious) case of Schrödinger's cat, appear almost paradoxical; resolving these potential contradictions threatens even the reductionist principle itself – hence its inclusion in this book. I have written about quantum physics and its problems elsewhere (see Further Reading) so in this chapter I am going to concentrate on these aspects of the subject that appear to challenge reductionism.

Because many of the ideas of quantum physics are unfamiliar and counter-intuitive, some otherwise intelligent people seem to think that the subject is beyond their powers of comprehension. If understanding means providing an explanation in familiar, everyday language, this is not just difficult but impossible! By choosing appropriate examples, however, the fundamental ideas of quantum physics can be appreciated and understood without the need for convoluted argument or complex mathematical reasoning.

Particles and waves

In Chapter 2, we considered wave-particle duality through the properties of light: a beam of light behaves as either a wave or a stream of particles (photons). The greater the intensity of the wave (that is, the brighter the light) the higher the density of photons in the beam. Suppose that such a light beam is directed on to a 'detector', which produces an electrical signal proportional to the intensity of the light striking it. If the light beam is very intense, the signal from the detector will appear to have a constant value but if the intensity of the light is sufficiently reduced, the detector output consists of a series of pulses, each of which contains the same amount, or *quantum*, of energy, corresponding to the energy carried by a single photon. As the light intensity is further reduced, the photon pulses become rarer, although the energy carried by each does not change. Most of the discussion in this chapter will be based on the properties of individual photons and their interaction with large-scale objects designed to detect them, because this is where the conceptual problems are most marked.

Many of the basic principles of quantum physics are exhibited in experiments conducted using what is called a *partial reflector*.

This can be thought of as an imperfect mirror where, although some of the light falling on it is reflected in the same way as an ordinary mirror, the rest passes through as if it were transparent. We can see this in a piece of glass, such as a window; under suitable lighting conditions, we see our own reflection at the same time as the scene on the other side of the glass. In this case, only a fraction of the light is reflected but a more general partial reflector can be constructed by applying a thin layer of silver to a piece of optical glass. The fraction of light transmitted depends on the thickness of the silver layer; it is typically 50 percent, in which case the mirror is often described as 'semi-silvered'.

Consider how the light would be expected to behave if its properties were those of a wave. The light wave striking the semi-silvered mirror would be split into a reflected part and a transmitted part, as in Figure 8.1(a). The intensities of the light in the two beams could be measured by placing a detector in each; the ratio of these intensities would be determined by the amount of silvering. This is indeed what is observed when the intensity of the light is too great to allow observation of the individual photons.

Now suppose that the light striking the mirror in Figure 8.1(a) is of such low intensity that the individual photons are detected. We find that each detector records a series of pulses, just like those discussed earlier. The energy in each pulse is the same as it was before. The reflector has apparently not caused the photons to split but has directed each one into either the transmitted or the reflected beam. The average rate of arrival of photons in each detector is half that observed for the full beam and each is detected in one or other of the beams at random.

A somewhat different set-up is shown in Figure 8.1(b). Two fully reflecting mirrors have been introduced, which have the effect of bringing the two beams together on the surface of a second semi-silvered mirror. The detectors have also been removed from their previous positions and placed so as to detect the photons in the beams that emerge from this second semi-silvered mirror.

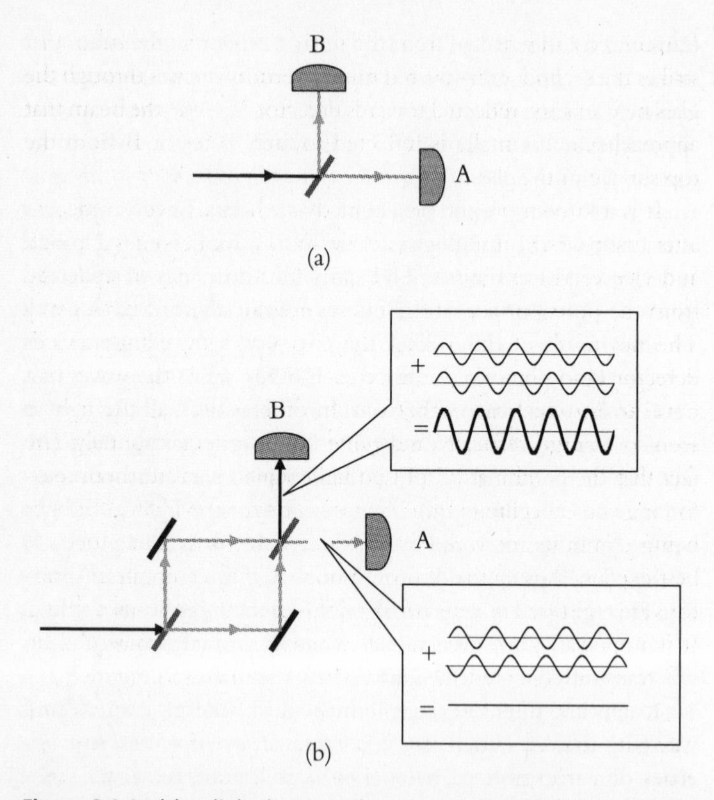

Figure 8.1 In (a) a light beam strikes a semi-silvered mirror and the transmitted and reflected beams are directed onto detectors marked A and B respectively. In (b) the light beams leaving the semi-silvered mirror first strike fully reflecting mirrors, which then direct them onto a second semi-silvered mirror, before being detected. In the latter case the waves recombine and are reinforced in the emerging vertical wave, but cancel each other out in the horizontal direction. (Alastair Rae)

In this set-up, the wave model of light implies that the light wave divides into two parts, which reunite at the second semi-silvered mirror. The light beams that emerge towards detectors A and B are then both formed by combining beams that have been transmitted and reflected from the mirror. Note that the beam that strikes the second semi-silvered mirror vertically passes through the glass before being reflected towards detector A, while the beam that approaches horizontally is reflected towards detector B from the top surface of the glass.

It is a known property of light that when a wave is reflected after first passing through glass, what was a peak becomes a trough and vice versa but transmitted beams and those that are reflected from the outer surface of the mirror are not affected in this way. The net result of this is that the two waves travelling towards detector B combine to reinforce each other, while the waves that travel to A cancel each other out. In other words, all the light is seen to emerge vertically and none comes out horizontally. The fact that the combination of two light signals leads to them reinforcing and cancelling in this way is very strong evidence for light being composed of waves, even though the detection process is best explained using the photon model. In this set-up, all the photons are registered in detector B, while detector A records nothing. It is impossible, however, to tell whether any particular photon was transmitted or reflected at the first semi-silvered mirror.

It appears, therefore, that light behaves like two interfering waves up to the point where it is detected, when it has the properties of a stream of particles. This is still true even if the light intensity is very weak; in fact so weak that no more than one photon passes through the apparatus at any time. Although the rate of detection of the photons is very much less, the pattern observed is otherwise unaltered: all the photons emerge through B and none are detected by A. Consider what would happen if the second semi-silvered mirror was removed: the horizontal and vertical waves would cross each other but would not combine in the way they did before. Equal numbers of photons would be registered in each detector and we would know that each photon arriving at detector B must have been transmitted through the first semi-silvered mirror, while all those detected in A must have been reflected.

The strange thing about these results is that the properties of the light itself seem to be determined by the way the apparatus is arranged. If there is only one semi-silvered mirror, each photon is detected in one or other of the detectors and the path followed by each photon is known, but if there are two, the light appears to pass along both paths in the form of waves which combine at the second semi-silvered mirror; after this it is detected as a stream of particles. But the two set-ups are essentially the same up to and beyond the first semi-silvered mirror, so how do the photons 'know what they are going to do'?

The Copenhagen interpretation

The above scenario exemplifies the conceptual problems encountered in quantum physics. How these can be resolved remains a controversial question for the scientists and philosophers who work in this area. The first point to make is that no actual contradiction is implicit in what has been described so far. This is because the arrangements of the apparatus in both cases is different: in one case, the photons pass along one or other path at random and in the other, the light beam has apparently divided and passed along both paths at once. The results of both experiments are fully explained if the model used to describe the light depends on how the counters are positioned. This is still the conventional view, first set out by the Danish physicist Niels Bohr in the 1930s. Bohr was one of the pioneers of quantum physics and he took a particular interest in its philosophical implications. He founded a research institute in Copenhagen; the ideas have become known as the *Copenhagen interpretation*.

The general principle underlying the Copenhagen interpretation is that it is wrong to attribute properties to any quantum system unless they can actually be observed. We should not think of the wavefunctions that are calculated by the equations of quantum physics as direct representations of reality but as abstract mathematical entities used to calculate the probabilities of different possible results. Contact with reality is made only when a measurement is performed using large-scale laboratory apparatus, such as photon detectors. The type of phenomenon that emerges depends on the nature of the experimental set-up as a whole. From the Copenhagen interpretation's point of view, if the second semi-silvered mirror is in place it is meaningless to say that the photon has followed a particular path, because there is no way that this can be observed. We should accept that we cannot fully understand the quantum world (perhaps because humans have evolved with hardly any direct contact with it) and our descriptions of reality should take this into account. It is meaningless to attribute any reality to properties in a context where they are unobservable in principle.

Niels Bohr summarized this way of looking at quantum physics in the sentence: 'There is essentially the question of an influence on the very conditions that define the possible types of prediction regarding the future behaviour of the system'. In Figure 8.1, 'the very conditions' means the way the apparatus is arranged, in particular whether the second semi-silvered mirror is in place. The 'possible types of prediction' means which path a photon has followed if the semi-silvered mirror is absent and the fact that all the photons will be detected in the vertical detector if it is in place

A major objection to the Copenhagen interpretation is that it relies on the assumption that when light strikes a photon detector, either a photon is detected or it is not. But if quantum physics is a universal theory, it should be just as applicable to the photon detectors as to anything else, so should not the detectors themselves have associated wavefunctions subject to the Schrödinger equation? Counters are immensely complex objects, composed of many atoms, each containing a number of electrons. Their wavefunctions will also be correspondingly complex; a detailed quantum analysis of this situation is almost certainly impossible in practice. Nevertheless, it can be shown that the fundamental properties of the Schrödinger equation predict that in the set-up illustrated in Figure 8.1(a) (or Figure 8.1(b) with the second semi-silvered mirror absent) the wavefunction associated with the detectors should be a combination, or *superposition*, of one part representing a state where a photon has been detected in the horizontal and not the vertical beam, and another part representing a state where it has been detected in the vertical but not the horizontal channel. In other words, the Schrödinger equation implies that this experiment should have no definite outcome in terms of the photon being detected in one channel or the other: both possibilities should be included in the final quantum state.

Erwin Schrödinger himself proposed a graphic illustration of the above behaviour in the 1930s. He suggested a thought experiment in which the signal from one of the particle detectors, in a set up similar to that shown in Figure 8.1(a), would trigger a lethal device directed at an unfortunate cat. Following the same logic as in the previous paragraph, a direct application of the Schrödinger equation predicts an outcome where the cat is neither alive nor dead but in some combination of both! Animal lovers should be assured that, as far as I know, this experiment has never been performed but if it were to be, we would be confident that the outcome would be either a dead cat or a live one and not one in suspended animation!

To ensure that quantum physics agrees with experiment, an additional principle has to be added to the picture derived from the Schrödinger equation. This is sometimes referred to as *wavefunction collapse*. As a result of a particle being detected, the wavefunction representing the particle, along with the detector, 'collapses' from one representing a superposition of possible outcomes into one corresponding to the actual outcome of the measurement. This means that after the measurement Schrödinger's cat is *either* alive *or* dead and not in some kind of suspended animation. It is important to note that wavefunction collapse is an additional assumption required for the quantum predictions to agree with experimental observations.

An implication of the above is that the physical laws controlling the properties of objects like detectors and cats are different from those governing elementary particles like photons, because wavefunction collapse occurs in the first case but not the second. This has become known as the *quantum measurement problem*. Why is it a big deal? One reason is that it at least appears to constitute a breach of the principle of reductionism: large-scale objects obey laws that are fundamentally different from those applying to their component particles. We saw in Chapter 2 that the basic principles of quantum physics successfully account for the behaviour of photons, electrons, atoms, molecules, and so on but it seems now that they predict behaviour for counters and cats that is different from what is actually observed. Given this, the principle of reductionism seems to be breached.

Despite all of this, there is much that can be predicted with confidence by quantum physics. Although exactly when a detector will next record the arrival of a photon cannot be calculated, the probability of this occurrence and therefore what the average behaviour of a light beam consisting of a large number of photons can be. In earlier chapters, when I discussed the physics of atoms and molecules, we saw, for example, that the properties of ice depend on the electrical interactions between the component molecules, which result from the fact that the electrons are more likely to be found on the oxygen rather than the hydrogen atoms. In this and similar cases, however, it is the *average* charge distribution that is important – the detailed quantum behaviour of the individual electrons is essentially irrelevant. Similar considerations apply to the organic and biological molecules discussed in Chapters 4 and 5: their properties can be understood on the basis of the average behaviour of the component electrons as described by quantum physics and the difficulties discussed above have little, if any, relevance.

In summary, in nearly all practical cases the physical properties of large-scale objects are determined by the average behaviour of their component particles, even though the latter are subject to the laws of quantum physics. To this extent, there is no challenge to the principles of reductionism: the laws governing the average behaviour of their fundamental particles completely determine the properties of large-scale objects. This, however, avoids rather than resolves the problems discussed in this chapter. When particles individually interact with counters or cats, wavefunction collapse occurs and the objects exhibit behaviour quite different from that apparently predicted by the laws of quantum physics as reflected in the Schrödinger equation. This may well constitute the biggest challenge faced by reductionism and there are some scientists and philosophers who believe the existence of these phenomena falsifies the reductionist principle. The originators of the Copenhagen interpretation, however, would not have agreed with this: they saw quantum physics not as a direct representation of the physical world but as a means of calculating the likely outcomes of experiments.

Beyond Copenhagen

Many have found the Copenhagen interpretation unsatisfactory and have looked for an alternative that would include a more realistic description of nature at all levels. Quite early in the development of quantum physics, various thinkers became dissatisfied with the standard approach. These included Albert Einstein, who believed that the aim of physics must be to provide a 'realistic' description of nature. By 'realistic', he meant that the properties of an object should be independent of how it is being observed, whether by an actual human observer or as a result of the experimental arrangement. Other scientists took a similar view and over some years an alternative view, now known as the *De Broglie–Bohm interpretation*, was developed. Prince Louis de Broglie was one of the pioneers of quantum physics, having originated the idea that beams of particles, such as electrons, could also display wave properties (see Chapter 2), while David Bohm was a physicist who worked in America until he fell foul of the McCarthy witch-hunt in the 1950s; he then moved to London, where he worked until his death in 1992. The basis of the De Broglie–Bohm approach is simply to say that quantum systems are composed of both particles and waves at all times and both of these really exist. They proposed that the role of the wave is to help determine the motion of the particle – that is, to guide it as it moves through space. In the example illustrated in Figure 8.1(a), they assume that a real, physical wave passes along both paths through the apparatus. When a particle enters the partial reflector, the wave guides it into one path or the other, where it is detected in the corresponding detectors. Which path a particular particle follows depends on where it is in the beam when it enters the reflector as well as the intensity of the wave in each path. This model then predicts the relative probabilities of the particle appearing in one or other detector and these turn out to be just the same as predicted by the standard collapse interpretation. Moreover, if the set-up is changed to correspond with Figure 8.1(b), the real physical wave also changes and it can be shown that it now guides all the particles into the same detector.

The De Broglie–Bohm interpretation may seem quite obvious and, indeed, it corresponds to the picture many people form when they are introduced to quantum phenomena: wave-particle duality is interpreted literally, as meaning that both entities are always there. Moreover, there is no challenge to reductionism, because higher level phenomena are fully determined by this lower-level substructure. Despite all this, however, the De Broglie–Bohm interpretation has only a few adherents.

The reasons for this are quite technical but the main points can be summarized. First, if the particle and the wave are both assumed to be real, what is the function of the 'empty wave' – that is, the wave occupying the path that is not followed by the particle? Second, to postulate the very existence of both particles and waves when they cannot both be observed seems like an unnecessary multiplication of entities that should be rejected by Occam's razor. The third objection relates to systems made up of two or more particles: in some situations it can be shown that the De Broglie–Bohm model implies that the particles must exchange information at a rate faster than light can travel between them. The impossibility of faster-than-light communication is a fundamental principle of Einstein's theory of relativity, which few physicists would be prepared to abandon.

Another approach to understanding quantum measurement is known as the *many worlds* interpretation. This was first put forward in the 1950s by Hugh Everett III, an American physicist and mathematician working at Princeton University in the US. He explored the consequences of assuming that the Schrödinger equation is always valid – that is, there is no wavefunction collapse. Referring again to the set-up described in Figure 8.1(a), Everett's assumption implies that the detector will be put into a state that is a combination of one in which it has detected a photon and another where it has not. If a cat were included, it would be in a state that is a combination of being alive and being dead. Instead of rejecting this out of hand, Everett realized that one consequence of the equations of quantum physics is that, once detection has occurred, there can be no communication between the alternative outcomes. The copy of the detector that has recorded the particle is completely unaffected by the copy that has not; the live cat is completely unaffected by, and is completely unaware of, the existence of the dead cat. This is because although we can bring together two light beams in a partially reflecting mirror – as in Figure 8.1(b) – it is completely impossible to perform the equivalent experiment with a counter or a cat. Only if we were able to ensure that every atom in these objects was in exactly the same state after the experiment as it was before could interference be demonstrated. As there are around 10^{22} particles in a typical counter, or cat, this is completely impractical. Moreover, it is impossible to isolate such large objects from their environment, so all the atoms in everything with which they interact would also have to be restored to their original state. The reason why this approach to quantum measurement is known as the many worlds interpretation is that every time a particle is detected, the 'world' – or indeed the whole universe – splits into branches, each of which is completely unaware of the other's existence. An important implication of the model is that the splitting also includes human beings: anyone who observes the result of the counter or cat will branch into one copy of themselves that has seen the cat die and another who saw it survive. But neither copy can ever be aware of the existence of the other.

Although the many worlds approach is termed an 'interpretation', it actually claims that taking the equations of quantum physics literally successfully predicts all our experience of the physical world, albeit at the cost of also predicting the existence of other parallel universes that we can never have any direct knowledge about. It presents no challenge to the reductionist principle, because the same laws control the physical behaviour of all systems, including photons, counters, and cats.

There are two main reasons why the majority of thinkers about quantum physics are less than enamoured of the many worlds approach. The first is pretty obvious: why should we believe in the existence of some huge number of parallel universes that we do not see? Isn't this a much greater violation of the principle of Occam's razor than either the idea that reductionism does not apply in the quantum context or that Bohm particles really exist? In one sense, this is obviously true; Everett's approach has been described as 'extravagant with universes'. However, this is not the way that Occam's razor is normally applied. This principle is that 'entities should not be unnecessarily multiplied', which means that we should not make any more *assumptions* than are needed to explain the observed results. Because Everett assumes only the fundamental equations of quantum physics (that is, the Schrödinger equation) it has been termed 'economical with postulates'. Whether or not we accept it then becomes a subjective judgement about the relative importance of economy and extravagance.

There is, however, another problem with the many worlds approach. A full description of the operation of a partial reflector requires not just the existence of two possible outcomes but also the fact that these may have different weights. That is (except for the special case where the reflector produces an exactly 50–50 split), the probabilities of detecting a particle in one channel or the other are not the same. But Everett claims that both experimental outcomes coexist and it is difficult to see how the concept of probability can make much sense in this context. Consider a two-horse race, in which you believe that one horse is more likely to win than the other. You might place a bet based on this judgement, hoping to profit if your favourite wins, although losing if it doesn't. But suppose you knew that at the end of the race everything, including the horses and you, was going to split into two branches, in one of which your horse won, while in the other it didn't. How could it then be meaningful to assign a probability to one outcome happening and the other not, when in fact both occur? How could you decide which future version of yourself to favour and therefore how best to place your bet?

The interpretation of quantum physics is still an area of considerable debate and controversy but it is important to realize that all interpretations agree about the outcomes of actual physical measurements. Whether it is assumed that the wavefunction collapses, the Bohm particle is detected or the universe branches, the probabilities of the results of actual observations are the same. Indeed, all the ideas set out in previous chapters are unaffected.

One example that illustrates this point is the effect quantum uncertainty might have on arguments related to free will, which I discussed in Chapter 6. It has been suggested that quantum physics may affect some brain processes so as to make their outcomes unpredictable and perhaps open to influence by human choice. The biochemistry of the neurons in the brain, however, is essentially deterministic, because it depends on the average behaviour of a large number of component atoms. Even if this were not the case, quantum physics would only introduce some randomness into their behaviour, with the probabilities of different outcomes being determined by the wavefunction. If these probabilities were to be modified by an outside influence so as to allow the exercise of 'free will' by some kind of non-physical mind, the basic assumption of random collapse would be breached. This would be just as radical a breach of the laws of physics as that involved if the mind were to alter deterministic behaviour. There is an unfortunate tendency to suggest that, just because there are still some unanswered questions connected with the understanding of quantum physics, these must relate to other unsolved problems, such as the nature of consciousness.

9 Conclusions

The two main themes of this book are, first, that the properties and behaviour of a physical system are controlled by the fundamental laws that apply to its components and, second, that genuinely new phenomena often emerge that would have been very difficult or impossible to predict from our knowledge of the components alone.

Both principles have played a vital role in the development of scientific thought over the last few centuries. The first has given scientists the confidence to use knowledge about how physical objects behave in one context to understand how they might behave in another. An early example of this was the development of Newton's theory of gravitation, which we discussed in Chapter 1. Newton's principal insight was to realize that the laws that compel objects to fall towards the Earth's surface also control the motion of planets and satellites in their orbits. If such generalizations were not possible, new laws would have to be discovered every time a different aspect of the physical universe was studied. Instead of it being a natural consequence of the principles governing the behaviour of its component atoms, there would have to be a separate law that ensured the impossibility of flying pigs.

The second principle, which has been termed *emergence* or *supervenience*, is also essential. Without it, our ability to understand the complexity of the physical world we interact with would either be severely limited or non-existent. Without concepts such as solidity and liquidity, the only way to describe the properties of large collections of atoms would be to account for the individual behaviour of each of them – an essentially impossible task, given the huge numbers involved.

These two principles of reductionism provide scientists with the power and confidence to modify physical behaviour at one level, using our understanding of how it follows from the properties of its components one, or more, levels below. Thus, an understanding of how the properties of materials known as semiconductors depend sensitively on the effects of quite small amounts of impurities has enabled the development of the transistor and the silicon chip. Understanding the silicon chip has led to the development of computers, which in turn have been used to develop the software needed for practical application in the home and elsewhere. Another example is the understanding of the genetic code, discussed in Chapter 4. The emergent concept of the gene has allowed botanists to develop new strains of plants by selective breeding or direct genetic modification. And, finally, the development of nuclear energy required an understanding of how the properties of radioactive materials supervene on those of their component nuclei.

Throughout this book, I have concentrated on tests of reductionism, which, if one or more of them had failed, would have falsified this central principle. Much scientific research, however, accepts the reductionist principle and applies it to the understanding of physical phenomena. In this process, theoretical ideas may be falsified by experiment without throwing doubt on the principle itself. During the study of biology, a theory may be proposed that, when tested experimentally, is found to be false. This will lead the investigators to modify or replace their theory with another that is still consistent with the underlying chemistry and physics. There is sometimes debate as to whether this type of study is as fundamental or important as those aimed at extending our understanding of the fundamental laws of matter. I remember many years ago hearing a discussion between a senior physicist, whose interests were the properties of solids and liquids, and a leading member of the University of Birmingham's high-energy particle-physics group. The latter said something like: 'what we are trying to do is discover new laws' and the former replied 'so are we'. By this, he meant that finding rules that describe the properties of bulk matter is just as challenging and important as the allegedly more fundamental research. Lord Rutherford (one of the outstanding pioneers of experimental atomic physics) would have agreed with the high-energy physicist; he opined that 'all science is either physics' (today he would mean particle physics) 'or stamp collecting', clearly implying the lack of fundamental importance of the other sciences.

The vast majority of scientists working today may be 'stamp collecting' in Rutherford's sense but their efforts to understand condensed matter in its many forms, including those exhibited in biology, should not be so readily dismissed. The understanding of how the properties of DNA emerge from the application of the basic laws of chemistry to particular assemblages of atoms must surely count as one of the greatest of human achievements. Similarly, the continuing research into the relationship between the conscious mind and the neurophysiology of the brain is an attempt to understand one of the greatest questions of all.

The application of reductionism to human behaviour – particularly social behaviour – is more controversial. Here is a quote from a collection of articles by the philosopher Mary Midgley published recently under the title *The Myths We Live By*:

> The same reductive and atomistic picture now leads many enquirers to propose biochemical solutions to today's social and psychological problems, offering each citizen more and better Prozac rather than asking them what made them unhappy in the first place.

The implication of this seems to be that solutions to social problems can only be found at the level of society. Certainly, society presents many challenges which can only be addressed collectively through social and political action. I touched on some of these in Chapter 7. But this is surely not a universal principle that should exclude action at the level of the individual. Why should someone who has a mental illness be expected to suffer while we wait until society has sorted itself out? To take another example, an epidemic of influenza presents a real problem to society but this should not prevent the protection of individuals by vaccination or the treatment of their symptoms by medicines. Because society emerges from the interactions of individuals, it surely must be true that some problems are best addressed at the level of society as a whole, while others will more readily yield to the treatment of the individual members. In many cases, the most effective treatments may involve a combination of the two approaches.

Anti-reductionism

Reductionist ideas are still sometimes seen as controversial and have been criticized by some serious thinkers. Much of the criticism relates to their application to human thought and society. As was discussed in Chapter 5, the renowned philosopher of science, Karl Popper, believed that a materialist explanation of consciousness is not possible. He endorsed the view of his fellow author, John Eccles, that the conscious mind is real and non-physical and interacts with the material brain through 'open synapses'. Another sceptic is the Australian philosopher, David Chalmers, who argues that no physical account of consciousness is possible. He has developed an alternative, *naturalistic dualism*, which involves making the existence of human consciousness a fundamental postulate alongside the basic laws of physics.

Unless we were convinced of the 'necessity' to do so (and not many have been convinced by Chalmers's arguments) such a 'multiplication of entities' would be a clear breach of the principle of Occam's razor. In any case, if these or similar ideas were to be accepted, the universal applicability of the reductive principle would inevitably be falsified.

One writer who has trenchantly criticized reductionism – in particular its application to human society – is the philosopher Mary Midgley. She was born in 1919 and, before her retirement, did much of her work at the University of Newcastle in the UK. In the 1980s, she wrote an excoriating review of Richard Dawkins's book *The Selfish Gene*, a popular account of the science of genetics and the principles of Darwinian evolution. One of Midgley's better-known controversial statements is that 'evolution, then, is the creation myth of our age'. On the face of it, this could be interpreted as saying that evolution has no more claim to truth than any other creation story, be it the biblical account in the book of Genesis or the dream time of native Australians. Such an interpretation, however, assumes that the word 'myth' means something that is not literally true but this is not what Midgley is saying. She is mainly interested in the influence of such myths on the culture of those people who follow them and believes the literal truth or falsity is of little or no relevance. Hence (my emphasis):

> By telling us our origins it [a myth] shapes our views of what we are. It influences not just our thought but our feelings and actions too, in a way which goes far beyond its official function as a biological theory. *To call it a myth does not of course mean that it is a false story*. It means that it has great symbolic power, which is independent of its truth.

With this gloss, it is difficult to see why Midgley should be thought of as an opponent of Darwinian evolution or why she is such a critic of reductionism. This use of the word myth seems to me to be close to the concept of the 'meme' (see Chapter 7), which Midgley strongly attacked.

The Myths We Live By also includes this statement:

> The fashionable reductive pattern tells us that, in order to connect different families of concepts, we should arrange them in a linear sequence running from the superficial to the most fundamental and ending with the most fundamental group of all, namely physics . . . Nor is it clear how this pattern of a one-dimensional hierarchy could ever have been applied. It could only work if the relation between physics and chemistry . . . could be repeated again and again, not only for biology but beyond that to colonize other branches of thought, such as history, logic, law . . . mathematics and to translate them all eventually into physical terms.

The implication that this is wrong or impossible goes against much that I have argued for throughout this book, where I have tried to explain how just such a linear sequence exists. As I have continually emphasized, an essential property of this is the emergence of phenomena, which have reality and importance in their own right, at each level. Midgley apparently does not recognise that emergence is an essential part of reductionism. This view is confirmed by another quote:

> Propositions such as 'a human being is only £5 worth of chemicals' or 'consciousness is just the interactions of the neurons' have the attraction of seeming to make life simpler because they are simple in themselves. The difficulty only comes when we try to work out what they mean and connect them with the rest of the world.

Again, this reveals that Midgley has not properly appreciated the meaning of reductionism as it is generally understood and as

I have tried to explain it. No scientist I know of could have intended either of the statements she quotes (without attribution) to be taken literally. The key words here are 'only' and 'just'. Yes, human beings are made up from chemical molecules but all the real attributes that make up their humanity emerge from the complex interactions between them, which are as essential as the molecules themselves. Yes, reductionism implies that consciousness emerges from the interactions between neurons but in no way does this diminish its reality or importance.

Censorship

One point I have frequently stressed has been to identify what evidence either exists or would be needed to *falsify* the reductionist hypothesis, rather than confirm it. In other words, if reductionism is true, it defines what cannot happen rather than what can – and in this sense it acts as a censor. Thus, if living things did not exist, it would be difficult or impossible to use only the fundamental laws of physics to predict their existence (quite apart from the problem of whom or what would be doing the predicting!). It is possible, however, to say that there are some things that human beings are not able to do and which, indeed, are generally forbidden if reductionism is true. These include Popper's and Eccles's idea of open synapses communicating with a non-physical mind as well as some more commonly held beliefs, such as:

- *Telepathy*. The transmission of information from one human brain to another means that neuronal changes in one must have been caused by similar changes in the other. This cannot happen unless there is some form of physical communication.
- *Astrology*. The positions of planets at the time of one's birth cannot be used to predict one's future life experience.

- *Homeopathy*. Any healing effect obtained from drinking water cannot be affected by dissolving some substance in it and then diluting the solution to the point where no atom of the substance remains.

These all refer to what is known as the 'supernatural' or 'paranormal' – though some might be better described as 'superstitions'. If the occurrence of any were to be reliably established, it would falsify the reductionist postulate, because changes in people's brains would be caused by something other than the laws of science that apply to their physical components: this is why they are termed *super*natural. Anecdotal evidence, and some research, supports the possibility of supernatural phenomena but so many of the results have been found to be based on fraud, poor experimental procedures, or just wishful thinking that it is generally agreed that such extraordinary occurrences would require extraordinary evidence before they could be accepted. This has not yet happened.

A question sometimes asked is why many people are attracted to the possibility of supernatural phenomena and why some passionately believe in them. One reason may be that if such phenomena occurred, this could be interpreted as evidence that the physical universe interacts in some way with human consciousness and that there is some purpose behind it. This belief is a fundamental principle underlying most of the world's religions, where it is nearly always coupled to questions of human morality and some possibility of survival after death. It is not difficult to see how the latter would have provided human society with a strong evolutionary advantage: those groups that contain members (particularly young men) who believe that death is not the end may well prevail over others who are likely to be more cautious. Even in the present time, a recent chief of the general staff of the UK armed forces has said that he would find it difficult to send soldiers into battle if he did not believe in an afterlife.

Such an expectation also apparently motivates some supporters of a cause to act as suicide bombers.

The necessity of religious belief or the lack of it, to underpin personal and social morality is often the subject of intense debate. It is not always recognized that whatever the outcome of this discussion, it actually says nothing about the correctness or otherwise of the metaphysical beliefs held by the proponents.

One interesting, if hypothetical, question is how the scientific community would react if the extraordinary evidence needed to establish a phenomenon currently thought to be supernatural were to emerge. One possibility is that it would be recognized that this falsified reductionism, but another is that efforts would be made to explain the phenomenon while keeping the principle intact; perhaps by identifying some presently unknown type of physical process that acts at all levels but only significantly at the large scale. Thinking of this type underlies some of the discussion concerning the quantum measurement problem discussed in the last chapter.

A phenomenon that certainly plays a part in the assessment of the efficacy of homeopathy, and indeed other branches of so-called 'alternative medicine', is the *placebo effect*. This is observed when a patient's health improves after having been given a substance they believe is efficacious, even though it is completely neutral. How our beliefs can affect our bodies is not at all well understood but it is the subject of active research and few of those studying this field would believe a supernatural explanation is required.

Another example of a little understood phenomenon is *ball lightning*. For many hundreds of years, occasional reports have emerged of people seeing bright balls of light that move through the air and sometimes pass through closed windows and doors. This has always been associated with thunderstorms – hence the name. Considerable research effort has been devoted to understanding the nature and cause of ball lightning. Efforts have been made to generate it in a laboratory, so that it can be examined in detail and theoretical studies that attempt to explain its occurrence using the laws of electromagnetism have been performed. Neither of these approaches has been completely successful and ball lightning remains an active area of research. Why is this the case, when the possibility of the occurrence of supernatural phenomena is readily dismissed? There are two reasons for this. One is that the evidence for ball lightning, though largely anecdotal, is quite robust and the other is that it is generally expected that ball lightning will eventually be explained as an electromagnetic phenomenon, with no challenge to the known laws of physics or the reductionist principle.

Mary Midgley has shown no sign of belief or interest in the supernatural and it may be that she has recognized the function of reductionism as a censor, because she has also written: 'In our time, reduction overwhelmingly presents itself as purely negative, a mere exercise in logical hygiene, something as obviously necessary as throwing out the rubbish'.

Is this what reductionism is – a vacuum cleaner for sucking up non-scientific, superstitious ideas? This is certainly one of its functions but, as was discussed earlier in this chapter, the reductive postulate has also been a vital tool in the development of both scientific understanding and technological application during the last few hundred years.

Final words

I should like to finish on a more personal note, with a final example that illustrates the principles of reductionism and which played a major part in my own scientific career. One of my main scientific interests has been the properties of solids. About a hundred years ago (before my time!) a completely unexpected phenomenon was observed by a Dutch scientist, Kamerlingh Onnes, at the University of Leiden in Holland. He was studying the electrical properties of mercury and found that, when cooled to a very low temperature, this substance completely lost any ability to resist the flow of electric current. Passing a current through a normal conductor, such as the copper wire used in domestic electricity supplies, requires the application of a voltage; the ratio of this voltage to the magnitude of the current is the *electrical resistance*. Onnes showed that the resistance of mercury at a very low temperature is exactly equal to zero, so that electric currents can flow without resistance. Some other elements have also been found to lose their resistance at low temperatures; this phenomenon is now known as *superconductivity* and the materials that exhibit it as *superconductors*. One of the practical applications of superconductors is in the construction of large powerful magnets, such as those used in MRI scanning: because of the absence of electrical resistance, the currents needed to produce the large magnetic fields flow without generating heat.

Superconductivity is a prime example of an emergent phenomenon. Solid mercury is composed of atoms arranged in a crystalline pattern (like those discussed in Chapter 3) with some of the electrons free to move and conduct electrical current. At normal temperatures, the electrons collide with the atoms in the crystal, which gives rise to resistance. At sufficiently low temperatures, this resistance disappears and superconductivity emerges. At the time it was discovered, it was completely inexplicable; it was more than forty years before a theory was developed which showed how superconductivity was a natural and inevitable consequence of the application of quantum physics to the behaviour of the component atoms and electrons in the solid. This explanation is too complicated to set out fully here but, briefly, it depends on the fact that the interactions between the mobile electrons and the crystal lattice of atoms result in a weak attraction between the electrons, which causes them to form pairs. These pairs interact with each other in a subtle way and this allows them to move through the solid without resistance. This is another example of an emergent phenomenon that supervenes on the fundamental laws that govern the component electrons and nuclei.

In the mid-1980s, Johannes Bednorz and Alex Müller, two scientists working in Switzerland discovered a new class of superconductors, ceramic materials with quite complex crystal structures. Their main, and surprising, property is that they become superconducting at temperatures that are much higher than those required by more traditional superconductors. Until this development, no substance was known to superconduct at a temperature greater than about 25K (–248°C) but the equivalent temperature for some of the newly discovered materials is well over 100K (–173°C). Although this is still much colder than room temperature, such substances are now known as 'high-temperature superconductors'.

At that time, I was working with other scientists at the University of Birmingham studying some of the detailed properties of the conventional superconductors but when the new discovery became known, we (along with just about every other similar laboratory in the world) switched our attention to the new materials. Luckily, we already had an experimental set-up that needed very little modification to test whether the electrons in high-temperature superconductors paired up as they do in conventional materials. We found that this was indeed the case and were able to publish this result quite quickly. Although the initial excitement died down, research into high-temperature superconductivity has continued. Despite this, however, there is not yet a generally accepted theory of why these materials behave the way they do. It is generally believed that some process occurs that results in particularly strong binding within and between the electron pairs, which allows superconductivity to occur at a relatively high temperature but there is no consensus as to what the process is. One suggestion no one has made is that the fundamental laws governing the behaviour of the individual electrons and atoms in these materials need to be modified. High-temperature superconductivity is confidently believed to supervene on the known physics of its components, even though how it does so has yet to be discovered.

I have tried to show how reductionism is applied at every stage in the journey from the structure of the atom to the collective properties of human society. I do not believe that anything we have encountered has been successful in falsifying the fundamental reductive principle but there are certainly areas where a fully constructive account has yet to be devised. Chief among these are the question of how consciousness emerges as a property of the brain and how the quantum measurement problem can best be explained. I believe that these, along with the ongoing problems of society are the most challenging questions for scientists and philosophers in the twenty-first century.

I have deliberately not addressed the 'science versus religion' controversy directly, though I expect my views have become clear. Certainly, many religious beliefs at least appear to run counter to the reductive principle, particularly where it applies to consciousness: dualism and a form of survival myth play an important part of most if not all faiths. Many of the arguments in favour of religion are based on claims of its ethical benefits to both society and the individual. Nevertheless, some (though by no means all) of the manifestations of religion give rise to division and conflict rather than co-operation. Many on the Christian right, particularly in the US, deny the threat of global warning and oppose actions aimed at addressing this problem, while the violence associated with extreme minorities that claim to be associated with the Muslim faith is well known. This has contributed to the emergence of the *new atheism* movement, which blames many of the problems of society on religion; the implication being that if everyone abandoned religious belief, society would have fewer problems. This may be true but I don't believe it has been objectively demonstrated; too many societies based on atheism have turned out to be despotic and totalitarian, asserting their own ideology just as fiercely as many theocracies. Moreover, the argument that knowledge of truth will inevitably lead to improved moral behaviour by individuals and society is an assertion that is too often made, by both theists and atheists, without any supporting evidence. Both sides often seem to assert the (originally Christian) precept that 'you shall know the truth and the truth shall make you free' without seeing the need to provide any justification or evidence.

Nevertheless, I believe that our only hope of avoiding the threats we face from nature itself (particularly in the form of global warming) and our own nature (particularly our tendency to generate conflict) is to increase our understanding of both the physical world and the workings of society. The systematic application of reductionism, along with the three principles of falsification, simplicity, and emergence to the understanding of society and its relationship to the individuals that make it up, will be essential if we are going to make the progress necessary for the human race to survive and prosper. Whether the human species has evolved to be capable of doing it is another matter.

Further reading

I have been greatly helped by the material in the following list of books. I have also made considerable use of web-based articles that can be located using an Internet search engine. I have not included any references to academic journals.

The Logic of Scientific Discovery, Karl Popper, Routledge Classics (2002). First published in 1935, this is the ground-breaking work in which the principle of falsifiability was first presented.

Quantum Physics: A Beginner's Guide, Alastair Rae, Oneworld Publications (2005). This contains an introductory account of the basic ideas of quantum physics and their application to atoms and molecules, which is somewhat more detailed than that set out in Chapter 2.

Endurance: Shackleton's Incredible Voyage to the Antarctic, Alfred Lansing, Phoenix (2000). This historical account of Shackleton's expedition contains the opening quote for Chapter 3.

Physics of Ice, Victor F. Petrenko and Robert W. Whitworth, Oxford University Press (2002). A detailed account which some readers may find rather advanced in places.

How We Live and Why We Die: The Secret Lives of Cells, Lewis Wolpert, Faber and Faber (2010). I drew heavily on this book while developing chapters 4 and 5.

The Blind Watchmaker, Richard Dawkins, Penguin (2006). A clear, accessible and detailed account of evolution by natural selection.

Creation: The Origin of Life/The Future of Life, Adam Rutherford, Penguin (2013). A very readable discussion of evolution, which includes an up-to-date account of research into the origin of the first cell.

Consciousness Explained, Daniel C. Dennett, Back Bay Books (1992). A quite early attempt do what the title implies.

Quest for Consciousness: A Neurobiological Approach, Christof Koch, Roberts & Company (2004). A quite advanced treatment of the neurobiology of the brain and its connection to consciousness, including an up-to-date account of the research on neural correlates of consciousness.

Minds, Brains and Science: The 1984 Reith Lectures, John Searle, Harvard University Press (1992). A reprint of the series of BBC Reith Lectures, which includes Searle's account of the Chinese room.

The Self and Its Brain: An Argument for Interactionism, Karl Popper and John C. Eccles, Routledge (1984). Sets out the arguments for interactionism and the ghost in the machine, which are discussed in Chapter 6.

Mindfield: How Brain Science is Changing Our World, Lone Frank, Oneworld Publications (2009). An account of some of the modern work on the brain and consciousness, including a description of the Ultimatum game and the idea of neuroeconomics.

The Self Illusion: Why There is No 'You' Inside Your Head, Bruce Hood, Constable (2013). A modern account of the case for the idea expressed in the title.

Shadows of the Mind: A Search for the Missing Science of Consciousness, Roger Penrose, Oxford University Press (1994). One of a series of books by this author that set out his ideas on a wide range of fundamental problems. This volume contains a detailed account of the computer halting problem and its implication for the understanding of consciousness.

An Inquiry into the Nature and Causes of the Wealth of Nations, Adam Smith, Bantam Classics (2003). The classic work referred to and quoted from in Chapter 7.

The German Ideology: Including Theses on Feuerbach and an Introduction to the Critique of Political Economy, Karl Marx and Friedrich Engels, Prometheus Books (1998). One of a huge number of publications by these authors, which contains the passage quoted in Chapter 7.

The Frock-Coated Communist: The Revolutionary Life of Friedrich Engels: The Life and Times of the Original Champagne Socialist, Tristram Hunt, Penguin (2009). A highly readable account of Engels's life, which includes an accessible account of Marxist theories and how they were tested in the European revolutions of the nineteenth century.

Forecast: What Physics, Meteorology, and the Natural Sciences Can Teach Us About Economics, Mark Buchanan, Bloomsbury (2013). A recently published exposé of the inadequacy of the standard theory of free-market economics.

Quantum Physics: Illusion or Reality? Alastair Rae, Cambridge University Press (2004). A discussion of the conceptual questions raised by quantum physics.

The Myths We Live By, Mary Midgley, Routledge Classics (2011). The volume of collected works referred to and quoted from in Chapter 9.

Index